Great Ways to Learn Anatomy and Physiology

2nd edition

Charmaine McKissock

palgrave

First published 2009
Second edition 2014

First published by
PALGRAVE

Palgrave in the UK is an imprint of Macmillan Publishers Limited, registered in England, company number 785998, of 4 Crinan Street, London N1 9XW

Palgrave Macmillan in the US is a division of St Martin's Press LLC, 175 Fifth Avenue, New York, NY 10010.

Palgrave is the global imprint of the above companies and is represented throughout the world.

Palgrave® and Macmillan® are registered trademarks in the United States, the United Kingdom, Europe and other countries.

ISBN 978-1-137-41523-3 ISBN 978-1-137-41525-7 (eBook)
DOI 10.1007/978-1-137-41525-7

This book is printed on paper suitable for recycling and made from fully managed and sustained forest sources. Logging, pulping and manufacturing processes are expected to conform to the environmental regulations of the country of origin.

A catalogue record for this book is available from the British Library.

Great Ways to Learn Anatomy and Physiology

Contents

Acknowledgements

"Any intelligent fool can make things bigger, more complex, and more violent. It takes a touch of genius – and a lot of courage to move in the opposite direction."
E.F. Schumacher

"Simplicity is about subtracting the obvious and adding the meaningful."
John Maeda, *The Laws of Simplicity*, The MIT Press

This second edition of *Great Ways to Learn Anatomy & Physiology* is dedicated to the memory of Dr Soames Michelson. They broke the mould when they made this doctor. He could always find an original and humorous way to help any struggling learner understand and remember something.

As I wrote this book, I remembered many inspirational students and colleagues, as well as all those individuals who have been kind to me during this project and beyond. Thank you also, Bill Piggins, Phil Rathe, Mervyn Thomas, Linda Auld, Bryony Ross, Suzannah Burywood and all the Palgrave Macmillan team, for your brilliant contributions.

It's often difficult to know exactly where ideas come from: I apologize if you feel you have not been acknowledged. Many thanks to all those that have contributed in any way to this book: you know who you are...

Foreword

Dear reader,

Like many others, you have chosen this book to help you study Anatomy and Physiology (A&P) for career purposes. This is *not* just another A&P textbook to add to the growing pile. It is a special resource put together for you, based on the work of lecturers and students who know how tough it can be to study.

Understanding how bodies work (or don't) is fascinating, but remembering thousands of facts, numbers, spellings and diagrams can be very challenging; especially nowadays, when we are all bombarded by so much information that we have to absorb, understand, evaluate, store and eventually use.

As your time and energy are limited, we only present to you proven ways to make studying easier for yourself. You will be able to train your brain to function better, whilst relaxing your mind and enjoying learning more.

Basically, you already have what it takes to succeed in your goals:

- a **limited amount of time** – for trying out options to see what suits you best;
- an **open mind** – to add new ideas to the ones that already work for you;
- a desire to have a **relaxed mind and body** – this will help your brain to function much better;
- a **sense of humour** – it will keep you sane;
- some **sticking power** – the motivation, self-belief and determination to succeed;
- some **good A&P textbooks** and **human beings** to talk to are also essential.

We have organized this book around 10 chapters – based on the 10 Top Tips for learning that have already worked for countless students in your situation. Keep this resource close to you as you study, and dip in and out whenever you need it.
We hope that you will succeed with this unique approach to learning Anatomy and Physiology . . .

Introduction

The Memory Bottle

Imagine that all your old memories are stored in a bottle (your brain). It's a rather strange bottle, with a narrow neck (your 'short-term memory') and a big bottom (your 'long-term memory').

If you want to top up your old memories with new information, you will need to get it down into the bottom of the bottle.

Now it's easy enough to get the information into the top of the bottle, but it can get stuck in the bottle neck and quickly evaporate, if you don't shake it about in the *right way*, at the *right times*.

Once you have succeeded in getting your fresh new memories into the bottle, they begin to settle and mix with old memories. You should be able to open your memory bottle whenever you want, and get some information out again.

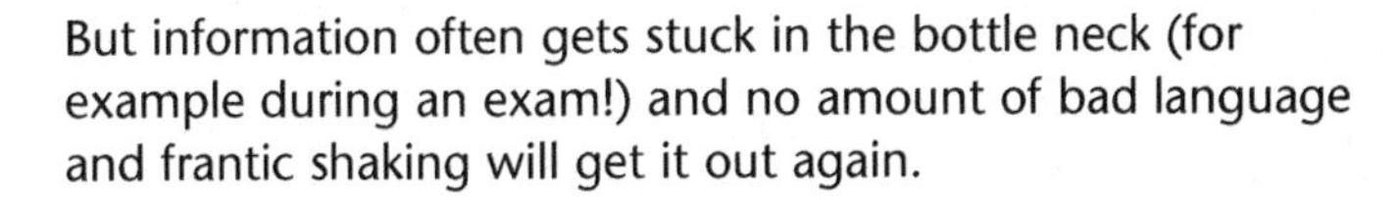

But information often gets stuck in the bottle neck (for example during an exam!) and no amount of bad language and frantic shaking will get it out again.

All 10 Top Tips in this book will help you get information in and out of your brain more easily.

10 Top Tips for easier learning: summary

1 Relax and take control

2 Bodymaps

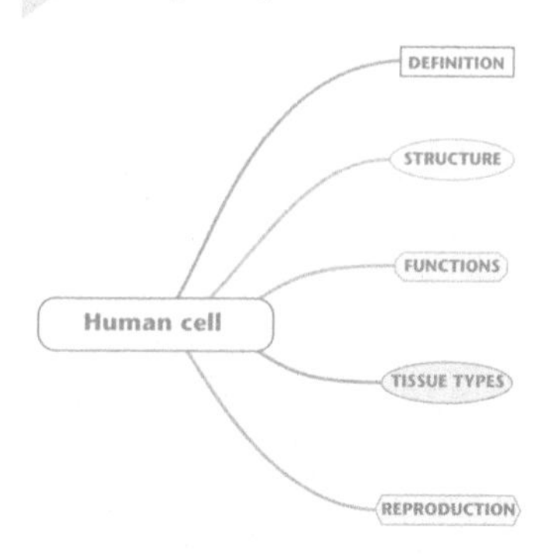

3 Timing is everything

4 A picture is worth 1000 words

5 Use all your senses

6 *Think-a-link*

7 *Memory tricks*

8 *Spelling: the 'WAM' way*

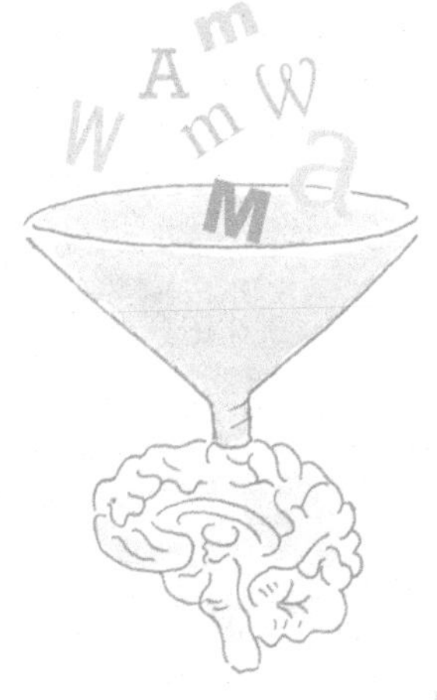

9 *Read right*

10 *Cool calculations*

Each tip has its own chapter in this book. You will find full explanations, examples and practice activities for each one over the next pages . . .

Do it your way!

The purpose of this book is to help you to learn A&P according to your *own* preferences – no matter how you are being taught. So, how do you choose which of the 10 Top Tips will work for you?

Before experimenting with each strategy, you may want to start by thinking a bit more about yourself. The more you know about yourself, the more you can use this knowledge to build on your strengths and get round some of your weaknesses. This will allow you to work 'smarter' not 'harder'. This strategy for success is based on the ideas of Howard Gardner – a famous professor of Education from Harvard University. He believed that '*It's not how smart you are that matters, what really counts is how you are smart.*' If you want to know more about the theory of 'Multiple Intelligences', read on . . .

Professor Gardner asked the famous question 'How would a Martian landing on Earth work out the intelligence of the human species?' Surely not through a set of paper and pencil IQ tests! This would be a big mistake, as it would certainly fail to identify humans performing exceptionally well in particular areas such as business, sport, music, art etc.

Gardner believed that all human beings have a wide variety of strengths and weaknesses for solving different kinds of problems. He suggested that there is not just one type of intelligence, but at least nine! These would include: Linguistic, Logical–Mathematical, Visual–Spatial, Bodily–Kinaesthetic, Interpersonal (Social), Intrapersonal, Musical, Naturalist and Spiritual intelligences. These are explained a little more on pp. 5–8.

Many academic subjects are still generally taught and examined in ways that mostly involve just two intelligences – your ability with words or with numbers. Good test results are not, however, good predictors of a happy, healthy, wealthy or successful life. Gardner believed that when you fully use all of your intelligences you really begin to release more brain power.

There are literally hundreds of 'Learning Style Questionnaires' that aim to help you decide which type of learner you are. But it is really important to know that most of us have a mixed bag of intelligences and it is a mistake to label someone as simply 'musical' or 'visual'. The following Questionnaire will help you explore your preferred learning style.

Questionnaire: What is your preferred learning style?

Work through the list of different 'intelligences' below, to help you decide which are your strongest and weakest points. This will allow you to focus on ensuring you make the most of your strengths and develop some of your weaker points. You may find it easier to get together with another person and interview each other.

Linguistic intelligence is the ability to communicate easily with words. Writers and politicians can be examples of people with good linguistic intelligence.

- ☐ Do you read a lot?
- ☐ Do you enjoy word games?
- ☐ Are you a good writer?
- ☐ Are you a good story-teller?
- ☐ Do you like to talk through problems, give your opinion, or ask questions?
- ☐ Do you easily absorb information from lectures, audio recordings or radio?

Logical–mathematical intelligence is the ability to reason, to make calculations, to plan and to think things through in a logical, step-by-step way. Lawyers, doctors, engineers, accountants and detectives might be examples of people with this kind of intelligence.

- ☐ Do you enjoy working with numbers or logical puzzles?
- ☐ Do you find it easy to make detailed plans for a holiday or event?
- ☐ Was science one of your favourite subjects at school?
- ☐ Can you find particular examples in support of a general point of view?
- ☐ Do you find it easy to take a step-by-step approach to problem-solving?
- ☐ Do you enjoy listing and grouping information?

Visual–spatial intelligence is the ability to think in pictures or imagine a finished process in your head. Engineers, fashion designers, inventors, architects, builders, artists, sculptors, photographers, sailors and town planners might be good examples of people with this kind of intelligence.

- ☐ Are you creative and imaginative?
- ☐ Are you good at reading maps and navigating?
- ☐ Do you enjoy jigsaw puzzles?
- ☐ Do you like making a drawing when taking notes or thinking through something?
- ☐ Can you easily take things apart and put them together again?
- ☐ Can you imagine how things look from a different perspective?
- ☐ Do you prefer to learn from pictures or videos?
- ☐ Do you have a good visual memory?

Practical (or bodily–kinaesthetic) intelligence is the ability to use your body skilfully to make or do things and present ideas – for example, dancers, actors, sports people, builders, surgeons etc. use practical intelligence.

- ☐ Do you use a lot of body language to express yourself?
- ☐ Are you a practical person, who enjoys 'do-it-yourself' or craftwork?
- ☐ Are you good at sports or physical exercise?
- ☐ Do you like to move about when thinking through problems?
- ☐ Do you learn best by watching and then 'doing' something?
- ☐ Do you need to physically handle something to really understand it?

Social (or interpersonal) intelligence is the ability to work well with others, to make friends easily, to be a good listener, to communicate understanding and feeling, and to identify another's aims and motivations. Social workers, teachers, therapists, religious leaders and sales people might be examples of people with this kind of intelligence.

- ☐ Do you enjoy working with other people?
- ☐ Do others often ask your advice?
- ☐ Do you enjoy going out and socializing?
- ☐ Do you value having close friends?
- ☐ Are you a good communicator?
- ☐ Do you like learning by talking things through with others?

Emotional (or intrapersonal) intelligence is the ability to understand yourself, to be aware of your feelings and manage them well, to reflect on your behaviour and motivation, to set goals.

- ☐ Do you often spend time thinking about the important issues in your life?
- ☐ Do you recognize your feelings and can you manage them effectively?
- ☐ Do you have a realistic understanding of your own strengths and weaknesses?
- ☐ Do you set yourself goals and make plans to achieve them?
- ☐ Have you had counselling or read self-improvement books?
- ☐ Are you aware of your preferred learning style when trying to learn something?

Musical intelligence is the ability to enjoy/make music, sing in key and remember tunes, and to keep rhythm.

- ☐ Is it easier for you to remember something you have heard rather than seen?
- ☐ Does tapping out a rhythm or listening to music help you learn better?
- ☐ Do you learn better with music in the background?

Naturalist intelligence is the ability to recognize, appreciate, and look after the natural world. Conservationists, farmers, gardeners, fishermen, explorers and zoologists might be examples of people with this kind of intelligence.

- ☐ Can you recognize and name many different types of plants and wildlife?
- ☐ Do you like animals? Gardening? Exploring the natural world?
- ☐ Are you interested in health or environmental issues?

Spiritual intelligence: this one is quite difficult to put into words. It might involve intuition, feelings of connection to something bigger than your own self, religious feelings, or an appreciation of non-rational or inexplicable events.

- ☐ Do you sometimes know something is true without being able to explain why?
- ☐ Are you able to learn better when your mind is peaceful?
- ☐ Are you able to put your own situation into a larger perspective?

What have you learnt about your own types of intelligence?
How might that knowledge help you learn more efficiently?
You will find, in this book, examples of many different ways of learning.
Test them out, and use what works best for YOU.

Chapter 1

Relax and take control

OVERVIEW

In this chapter, you will have the opportunity to explore how your state of mind can affect your learning ability, and how to use this information to your advantage.

You will find out:

- ✓ **how to understand your own stress;**
- ✓ **what the nervous system has to do with stress;**
- ✓ **what really makes us act the way we do;**
- ✓ **how positive and negative emotions alter brain function;**
- ✓ **how to let go of harmful feelings and unwanted thoughts;**
- ✓ **how to reduce exam nerves;**
- ✓ **some techniques to calm you down, such as Mindfulness meditation, body relaxation and visualization.**

Understanding your own stress

Do you constantly worry about the mountains of information that you need to absorb? Do you panic at the mere thought of your exams? Does your head start swimming as stress chemicals flood into your bloodstream? Does a nagging voice at the back of your mind keep putting you down? With constant practice, you can train yourself to slip into a positive relaxed state of mind, where you feel in control. The rest of this section covers lots of different stress-busting solutions that have really worked for other students. But first, it's important to fully explore your own stress . . .

What is stress?

It's the feeling – physical or mental – that comes when demands are made on you. The unpleasant effects of stress hit you when there's a gap between the demands made on you and your ability to cope with them.

How much stress can you take?

Every person reacts differently to challenges and changes: they are not necessarily bad for you, except if you start feeling really out of control. Anything that happens to you – whether winning the lottery or losing your job – can throw you off balance.

Can you cause your own stress?

External events that you can't control – like falling ill or losing somebody you love – can cause a lot of stress. But often we are our own worst enemy: the way we act or think about ourselves can be very damaging; for example, when:

- our standards are impossible to meet;
- we want approval from others all the time;
- we give others the power to control us;
- we blame external circumstances and others for the way we feel;
- we over-react and are inflexible;
- we always expect to fail . . .

Questionnaire: Symptoms of stress

Do you routinely suffer from any of the common symptoms of stress?

- Are you getting headaches?
- Do you often have a sore neck, shoulders and back?
- Do you continually feel run down or ill?
- Does your heart sometimes beat very hard and do you feel panicky?
- Do you sometimes feel dizzy, nauseous or breathless?
- Do you feel irritable and get angry easily?
- Do you feel anxious or depressed?
- Are you eating, drinking or smoking excessively?
- Do you keep putting off tasks and finding it hard to concentrate?
- Do you constantly feel tired and have sleep problems?
- Are you having problems managing your time and belongings?
- Do you lack confidence and have a lot of negative thoughts?

When you feel threatened. stress chemicals flood into your bloodstream. If this happens too often, you can get physical or mental problems. If you regularly suffer from any of the symptoms, first check them out with your GP. If any of your symptoms are stress related, there's lots you can do: you can reduce the demands made on you and increase your coping systems: both ways work. Many solutions are suggested in this book.

Stress and the Autonomic Nervous System (ANS): Summary

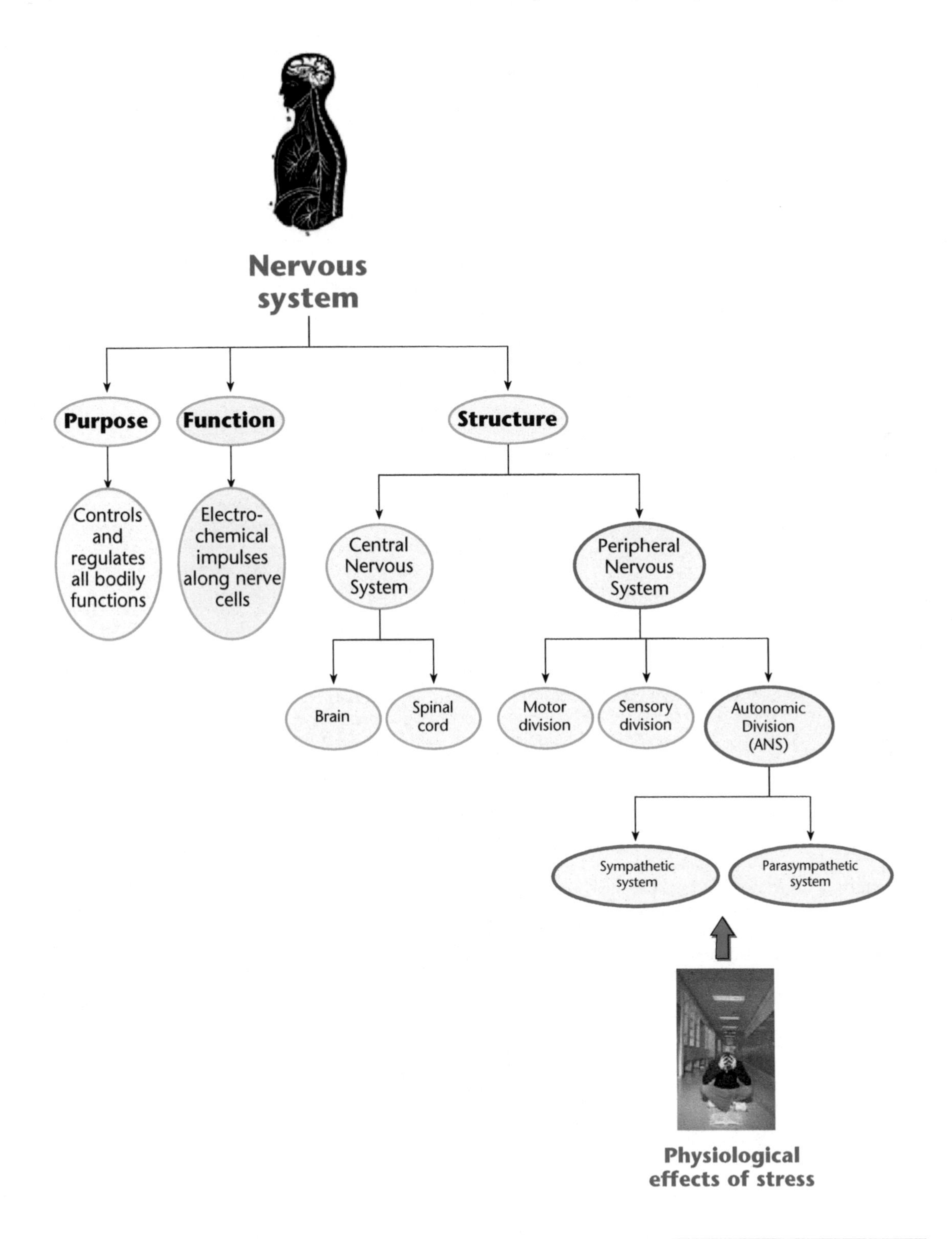

Diagram created in Inspiration® by Inspiration Software®, Inc.

What job does the ANS do in the body?

Your ANS is always working for you, without you knowing it, day or night. It keeps you functioning properly by regulating all involuntary functions (heartbeat, blood pressure, breathing, body temperature, hunger, thirst, sleep, emotions, water, sugar, salt levels, muscle tone, hormones, etc.). The process of maintaining a constant internal state despite external changes is called homeostasis.

The ANS keeps your body safe from immediate danger by preparing it for 'fight or flight' via the **Sympathetic Nervous System (SNS)**. When you feel threatened by something real or imagined, stress chemicals flood into your bloodstream.

The ANS then restores your body to a resting state via the **Parasympathetic Nervous System (PNS)**.

AUTONOMIC NERVOUS SYSTEM: fight or flight response

Structure	Effects of stimulated Sympathetic Nervous System	Effects of stimulated Parasympathetic Nervous System
Eye	Pupils dilate; eyes focus on distant objects (scanning for danger and escape route).	Pupils constrict; eyes focus on nearby objects.
Salivary glands	Saliva decreases: mouth gets dry.	Saliva increases.
Nose	Nasal glands produce less mucus.	Nasal glands produce more mucus.
Heart	Heart rate and blood pressure increase: more oxygen and blood to muscles.	Heart rate and blood pressure decrease.
Lung	Bronchial muscles relax; blood vessels dilate; breathing is faster.	Bronchial muscles contract and blood vessels constrict: breathing slows.
Stomach & intestines	Less digestive juices produced. Movement of food slows down.	More digestive juices produced. Movement of food increases.
Liver	Glucose released.	Glucose stored.
Bladder & kidneys	Sphincter closed. Less urine output.	Sphincter relaxed. More urine produced.
Adrenal glands	Produce stress hormones adrenalin and noradrenalin. Immune and reproductive systems suppressed.	Body brought to rest. Immune and reproductive systems active.
Skin	Blood vessels constrict; hair stands on end; sweating.	Blood vessels dilate. Sweat pores close.

What is the structure of the Parasympathetic Nervous System (PNS)?

The organs of the body have *separate* sympathetic and a parasympathetic nerve supplies.

The **SNS** consists of ganglia (a collection of nerve cells) connected from the spinal cord – from the thoracic to the lumbar area – and branching off to the organs

The **PNS** is connected at the sacral and cranial areas and branches off to the organs via different neural pathways.

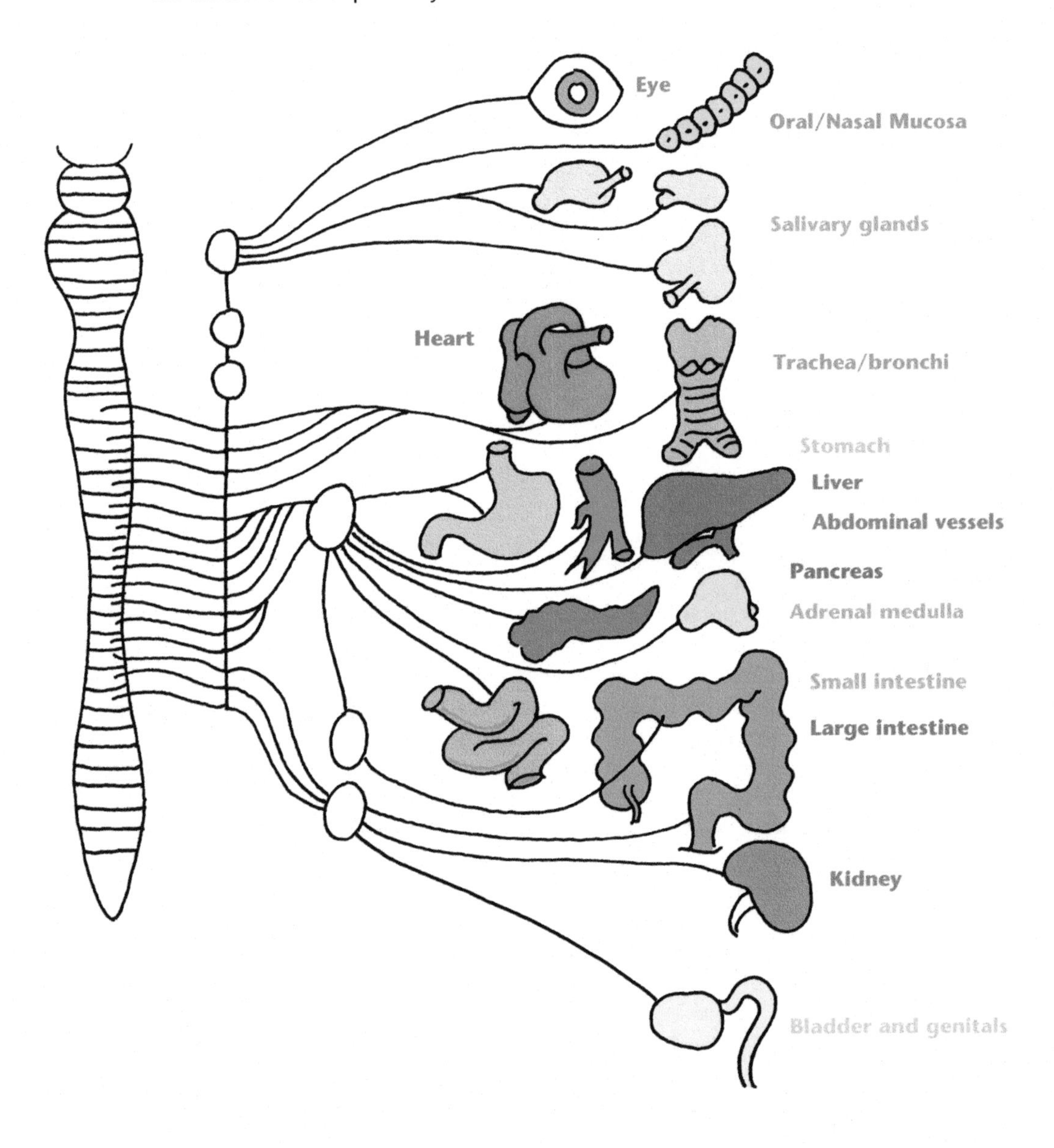

What really makes us act the way we do?

Most of us want to be better at something. It could be exams, relationships, dieting, exercise, time management, whatever! But the crazy fact is that we know exactly what we should do, but we don't do it. The reason is that we may be unaware of the strong physical, emotional and thinking patterns that affect our actions, often unknown to ourselves. Understanding our basic physiology can help us make changes that really last. For example, when you *think* 'I'm just rubbish at exams' you must first *feel* anxious about your own abilities. In order to *feel* anxious, the body's *physiology* must react with a rapid heart beat, irregular shallow breathing and tense muscles. Emotions feed thoughts and thoughts feed emotions in a continuous loop fuelled by your basic physiological systems: this circuit affects all your actions. Any attempt to just change our actions will inevitably be doomed to superficial and short-term changes.

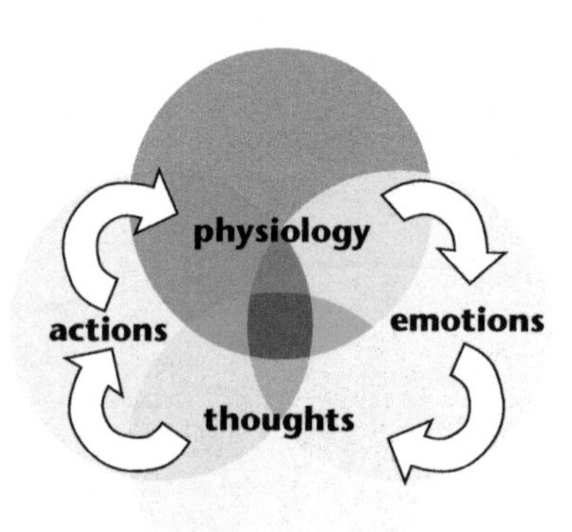

What drives your best performance?

One person may say that they perform best when they are aroused under pressure (in this state, adrenaline is released into the bloodstream). Another may need to feel relaxed (when acetylcholine is released into the bloodstream). Both are right and both are wrong: excessive arousal (fear/aggression) and excessive relaxation (apathy) will affect performance negatively. Dr Alan Watkins, a neuroscientist, has said that what really drives performance is whether the individual is in an *anabolic* or a *catabolic* state. Anabolic hormones (DHEA) are secreted when the individual is in a **positive** state of mind, and catabolic hormones (cortisol) are secreted when you are in a **negative** state of mind. The effect of high cortisol levels can be immense, ranging from weight gain, high cholesterol, depleted immune system, depression and, most importantly, impaired brain function. This is the reason why you may perform badly in exams when you are over-anxious, angry or lacking motivation.

You'll give your best performance when you are aroused with positive emotions or relaxed with positive emotions. Look at the following pages to explore your positive and negative emotions.

There are over 3000 emotions in English. Can you think of any more **positive** emotions to add to this map?

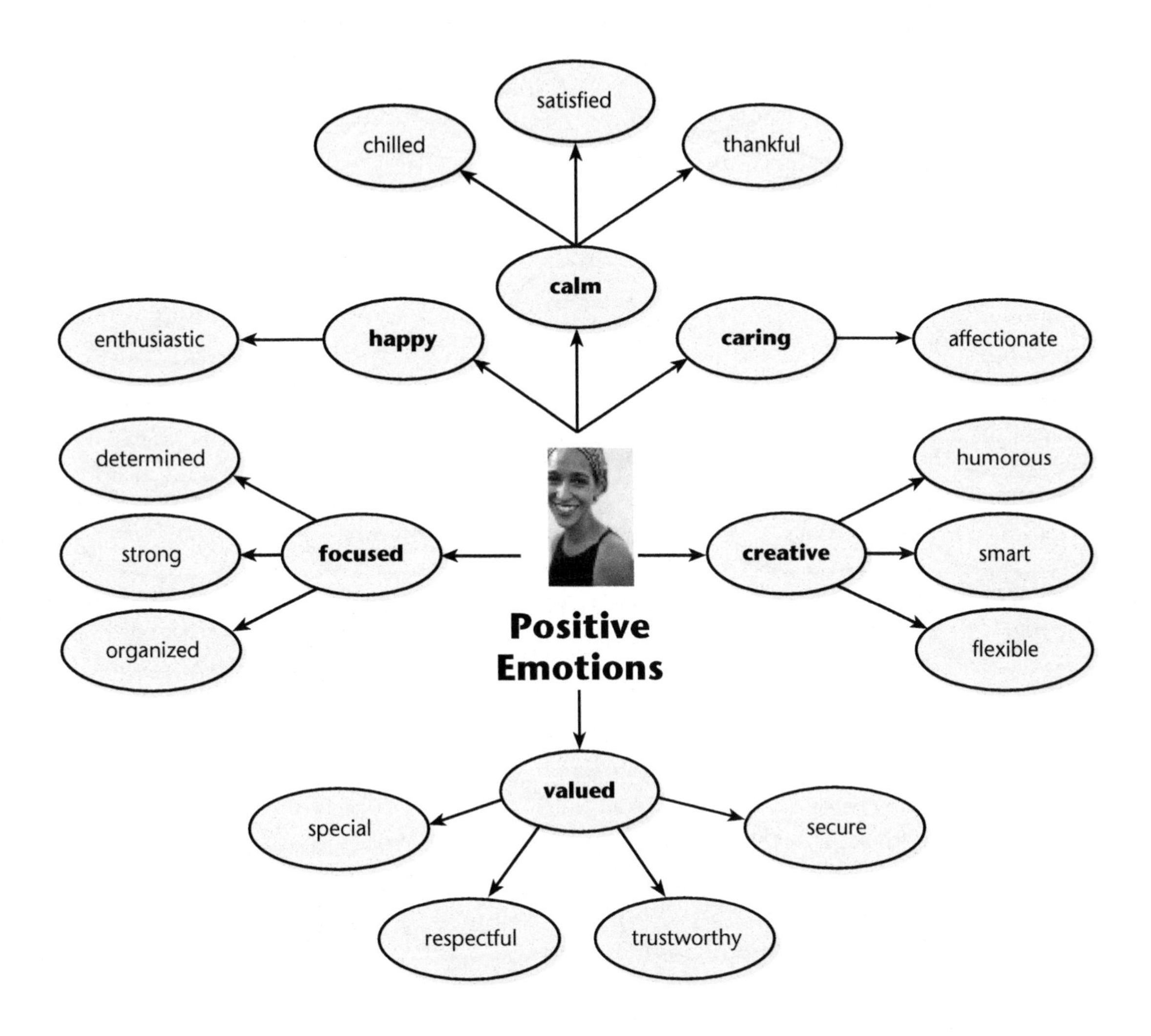

Remember: Positive emotions mean good chemicals flowing round your body and boosting your brainpower.

Diagram created in Inspiration® by Inspiration Software®, Inc.

Can you think of any more **negative** emotions to add to this map?

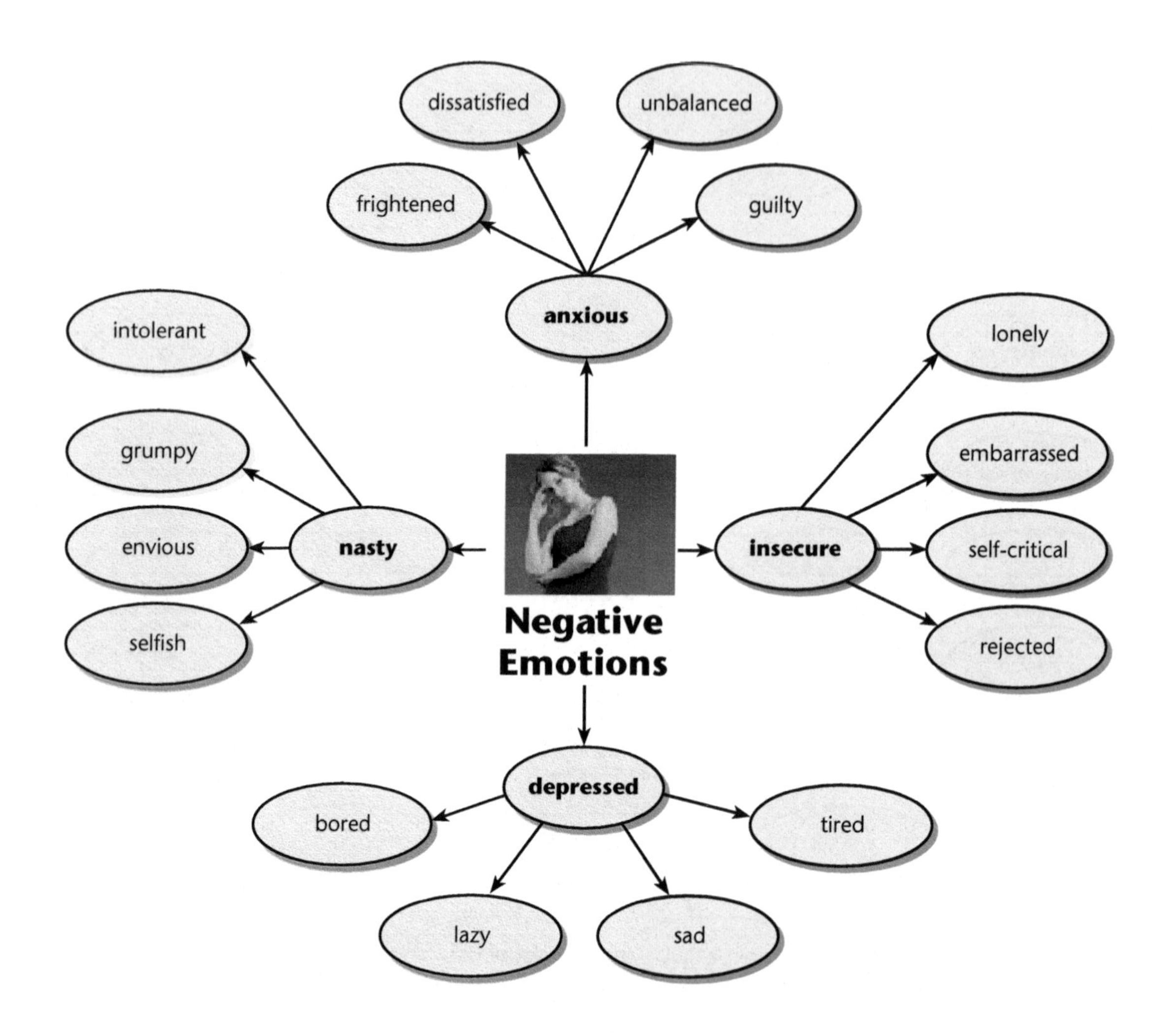

Remember: Negative emotions mean harmful chemicals flowing round your body and decreasing your brainpower.

Diagram created in Inspiration® by Inspiration Software®, Inc.

How do your emotions alter brain function?

The **Limbic System** includes several brain areas that are in charge of controlling emotions. These areas include the Cingulate cortex, the Hypothalamus, the Hippocampus and the Amygdala. The Amygdala is involved in memory, emotion and fear. The Hippocampus is involved in the forming and storing of memories. This shows the important relationship between memory and emotion. Strong emotions bring back memories and strong memories bring about emotions.

However, when emotions become extreme, your primitive instinctive brain area takes over and thinking processes become impaired. Your heart beats in an *erratic* manner, causing learning blocks. This kind of heart beat, where your heart signal goes up and down erratically, is associated with a catabolic state of mind. In this state, your **Frontal cortex** ceases to function effectively. The Frontal cortex is used for working-memory, time, sequence, planning, organization and attention shifting. This is why you get impaired decision-making, problem-solving, creativity and energy reserves. On the other hand, a *coherent* heart beat rate – one which is smooth and rhythmic – improves your thinking processes.

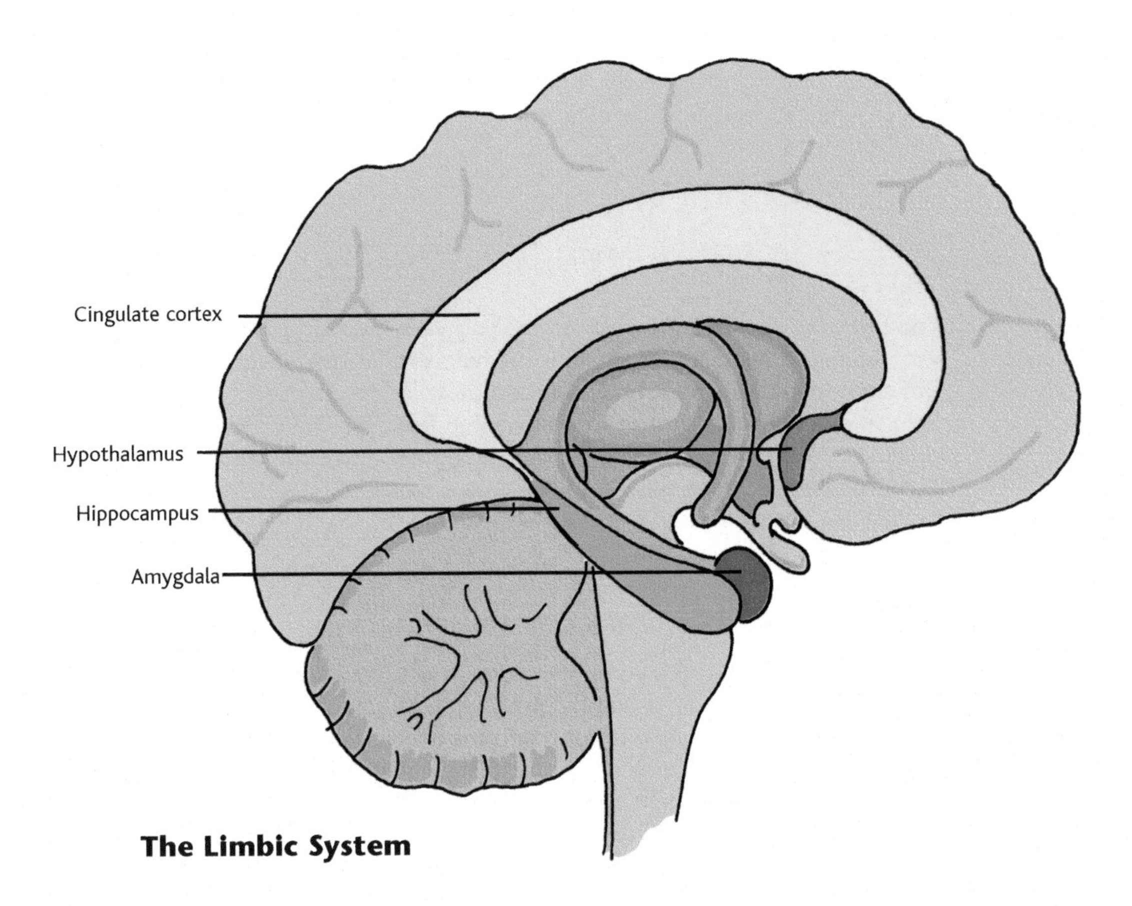

The Limbic System

What is the fastest way to boost brain function?

Dr Alan Watkins, neuroscientist and expert in Health & Performance, has researched the link between your breathing and your performance in situations such as study. When we get stressed, our breathing becomes erratic, irregular, fast and jerky. Look at the diagram below which illustrates the rapid chain of reactions in your body, when you feel threatened:

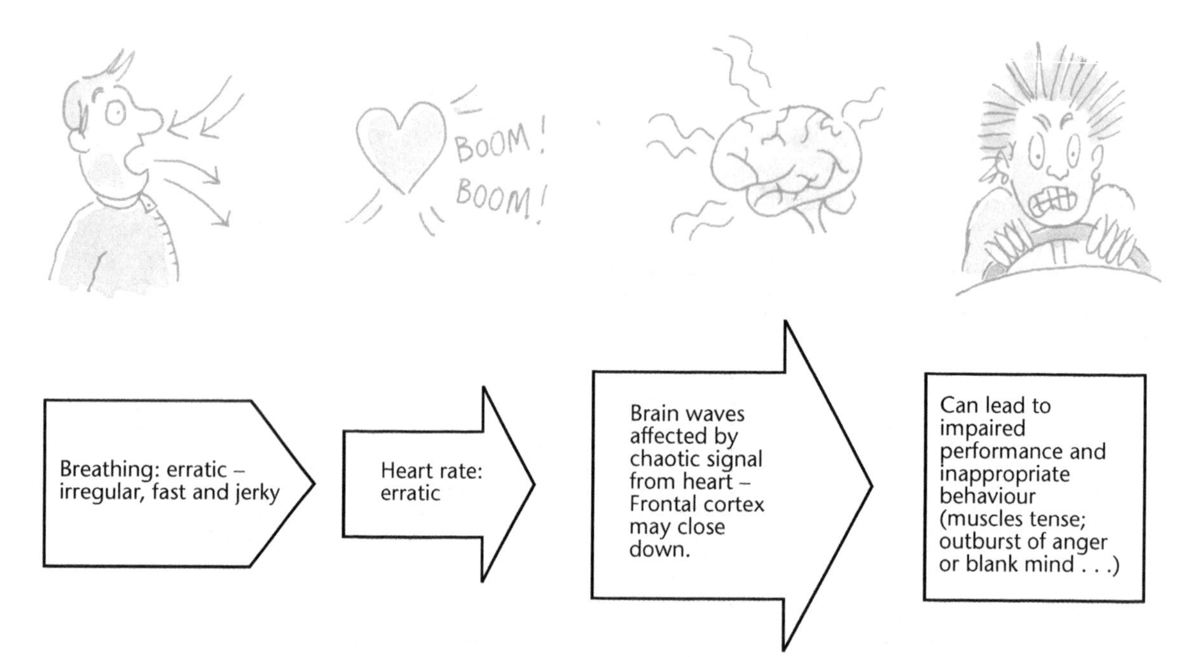

You can calm your heart beat by controlling your breathing. **Rhythmic breathing** is a very important life skill to learn, as it will help you keep your cool in challenging situations. It will stop your brain shutting down, allow you to maintain energy and help you identify and change your unwanted feelings. At first, you will need to practise making your breathing **rhythmic, regular and smooth**. This practice and other effective techniques, such as relaxation, visualization and Mindfulness meditation, are explained over the next few pages.

Controlling unwanted feelings

The ability to change your emotional states *when you want to* is another extremely useful life and study skill. It isn't helpful to wait till somebody or something else changes the way you feel: this just reinforces a feeling of powerlessness and dependency. Many people turn to quick fixes – such as alcohol, shopping, food, offloading feelings on others, etc. – to shift the way they feel. These strategies may be pleasurable, but they're not always available in the heat of the moment, their effects can be short-lived and they can often leave you feeling worse than before. A technique called Mindfulness meditation can help you control unwanted feelings, thoughts and actions: it costs nothing, is safe, and can be instantly available anywhere you go.

Clinical research about Mindfulness meditation

In Western psychology, 'Mindfulness' meditation is currently topical. It's a fresh approach to dealing with painful feelings and unhelpful thought processes. It's a powerful, evidence-based tool for reducing stress, increasing self-understanding and developing concentration. In fact, meditation can be as effective as medication to ease anxiety, depression, sleeplessness and pain. National Health and educational establishments have begun to offer short Mindfulness courses, and thousands of students are reporting great benefits.

How long does it take to start seeing benefits?

Even one session can leave you feeling refreshed and clear minded. Scans have shown that brain activity begins to change within just a few days of learning to meditate. Andy Puddicombe – former Buddhist monk and author of *Get Some Headspace* – finds that it generally takes about ten days to get comfortable with the basic techniques. These are easy to learn, but to alter brain pathways, a regular commitment is needed. Even a daily practice of five minutes is better than a longer weekly session.

How do you practise Mindfulness meditation?

In a nutshell, meditation is the practice of intentionally focusing your whole attention gently and non-judgementally on the present moment. The focus of your attention can be your breathing, bodily sensations, thoughts, emotions, external objects or physical actions. It is the opposite of day-dreaming and multitasking, which can give you the illusion of being productive. Each time your mind begins to wander, harps back to the past or worries about the future (minds are very good at this!) you bring it gently back to the here and now. This may seem boring, difficult and pointless to begin with, but these negatives soon disappear. The words 'relaxation' and 'meditation' are often used interchangeably, but are different. Meditation can lead to relaxation, but when you meditate, relaxation isn't the goal, more a pleasant side effect. With meditation, you also achieve more clarity of thought, self-knowledge, concentration, kindness and wellbeing.

Mindfulness borrows some of its practices from Buddhism. However, you don't need any religious or spiritual beliefs to practise mediation. It is designed to fit in with your schedule and lifestyle: you do not need to wear any special clothes, chant or sit crossed legged on the floor!

On page 21 you will find a sequence to practise. However, to start with, it's much easier to use a recording to guide you. (See 'Useful Resources' section page 170).

Questionnaire: How do *you* cope with stress?

Does your lifestyle help you deal with stress? To find out, answer the questions below truthfully. Tick the columns on the right that apply to you.

Are you looking after yourself?	Always	Usually	Sometimes	Never
Do you get enough sleep?				
Do you do something to relax completely every day?				
Do you take some exercise at least twice a week?				
Do you eat 3 times a day, with at least one balanced meal a day (carbohydrates, proteins, greens and fruit)?				
Do you drink at least a litre of water a day?				
Do you try to drink tea, coffee, or cola drinks in moderation?				
Do you smoke less than five cigarettes a day?				
Do you talk to somebody you trust when you are worried or angry?				
Do you recognize your own good qualities and appreciate the good things in your life?				
Do you know your weaknesses, but without beating yourself up about them?				
Do you try to focus on the present without always thinking about the past or worrying about the future?				
Do you try to succeed in what you do, while being able to learn positively from your mistakes?				
Do you say 'no' without feeling guilty, when asked to do something you really don't need to do?				
Do you manage your money and space well?				
Do you manage your time and energy well?				
Do you set yourself short-term and long-term goals?				
Do you use as many study strategies as possible: e.g., technology, memory techniques, relaxation techniques, help from others, etc?				

If you have answered mostly 'always' or 'usually', you seem to be dealing well with stress and you're looking after yourself.

If you have answered 'sometimes' or 'never', your lifestyle may not be helping you deal with stress effectively. Don't beat yourself up about it: take action. You can help yourself straight away by trying out some of the ideas and techniques presented in this chapter.

Instant calmer

We get so used to living with tension, that we don't even know it's there. If your muscles are chronically tense, you'll find it hard to relax. If you mind is tense, you'll find it hard to study.

Below you will find a technique to practise for about 10 minutes every day, if you can. It includes some relaxation, rhythmic breathing and meditation practices.

You can also use this technique in an emergency, when you need to quickly calm down, relax your body and clear your mind. You can do it sitting, lying down or even standing up.

It helps to make a recording of the sequence or get somebody to read the words out to you very slowly until the routine becomes automatic. There are also pre-recorded versions of similar sequences recommended in the 'Useful Resources' section on page 170.

You may find that unwanted thoughts keep coming back into your mind: try not to get irritated or upset with yourself – it's natural for minds to wander. Just keep letting go of unwanted thoughts, gently, kindly and without forcing anything. You may need to do this over and over again! The important thing is to notice when your mind wonders and gently bring it back to your breath.

Make your body as comfortable as possible

Loosen any tight clothing; make sure you are warm enough and won't be disturbed, if possible. Start by gently closing your eyes.
Be aware of any sounds inside or outside the room, and then let go of them.
Feel your clothes against your body.
Feel the air around you against your skin.
Be aware of any tastes or smells.
Let your seat or floor take all your weight, as your body sinks down.

Allow your mind to go blank

Notice any feelings or thoughts going round and round in your head.
Let go of any thoughts of the past.
Free your mind of any worries about the future.
When a thought pops into your mind, don't worry – it's normal, that's what minds do...Don't get caught up in any thought.
Just watch it gently floating away like a cloud across a blue sky.

Focus on your breathing for a few moments

Take a deep breath, hold it, and then breathe out very slowly.
Now feel cool air coming in through your nostrils and out through your mouth.
Then feel your breath rising and falling with each breath into and out of your heart area.
Then feel your chest rising and falling with each breath in and out.

Allow your body to relax completely

Scan your body very slowly for any areas of tension or pain, starting at your feet, moving up through your legs, up into your buttocks, your stomach, your hands, chest, back, shoulders, neck, face and head ...
Now clench your hands – tight, tighter ... hold the tension for a moment ....
Now, as you breathe out, let your hands go soft and relax completely ...
Now, your hands and arms feel pleasantly light and completely relaxed.

Pull in your tummy and buttocks as tight as you can. Tight ... tighter ...
Now breathe out, as you let your tummy and buttocks go soft and relax ...
Your tummy and buttocks feels completely relaxed now.

Push your feet into the floor. Push as hard as you can for a few moments.
Now breathe out, as you let your feet flop and relax completely ...

Now lift your shoulders to your ears and squeeze them as hard as possible.
Hold it for a moment, and now let go of your shoulders and relax your neck and shoulders completely ... That's right. Really feel the difference between tension and relaxation.

Now screw up your face, screw up your eyes and clench your jaw ...
And now let go of all your body tension from head to toe;
enjoy the feeling as your whole body relaxes for a while.

Rhythmic breathing

Now focus on your breathing: breathe in and out very slowly.
Imagine breathing in calm and breathing out any tension.
Follow your breath in and out, for 10 breaths. Count each breath.
If you lose concentration, that's OK. Start counting again from 1...

Finishing off

Rest in this calm and relaxed feeling for a few moments.
Have a yawn, a stretch and then start moving your body slowly. Take it easy. Slowly open your eyes, ready to meet the outside world, feeling relaxed, energized and able to cope better with whatever crops up...

How can you reduce exam nerves?

Negative emotions and excessive stress can seriously affect memory and concentration. Here is an action plan:

1. First check your current stress levels: pages 10 and 20.
2. Check your beliefs about exams in the 'Exploring myths about exams' questionnaire on page 23.
3. Practise the Instant Calmer on pages 21-2
4. Practise the Exam Visualisation on pages 24–5
5 Check out what other students have found useful on page 26.

Questionnaire: Exploring myths about exams

Distorted worries about exams can interfere with your performance, as well as making you really miserable. Below is a series of questions to help you identify if you are expecting too much of yourself. Tick the statements you agree with, and then look at the answers at the end of the book.

1 a. Having a really good memory is all that counts in an exam. ☐

b. Some answers in Anatomy and Physiology do rely on remembering pure facts. But usually you'll get better marks if you can show good understanding of your subject. ☐

2 a. Examiners like to trick you into making mistakes. If you show any weaknesses, they'll try to fail you. It's not fair. ☐

b. Examiners want you to pass: they'll give you marks for making a reasonable attempt at key points. ☐

3 a. You should answer only the specific aspect of the question that is asked of you – without waffling and padding. ☐

b. In exam answers, you should put down at great speed everything you know about the subject in question. ☐

4 a. You can do a very quick brainstorm plan, but there's no time to write out a carefully crafted answer. ☐

b. For each question, you need to write a plan, then an introduction, a middle and a conclusion. ☐

5 a. You will be marked mostly for content, but you may lose some marks if your writing or diagrams are difficult to understand. ☐

b. The presentation of your work must be very neat and attractive, with perfect spelling, punctuation and grammar. ☐

6 a. As you get older, your prior knowledge and experience of the world can really help you learn new things. ☐

b. The older you get, the worse your memory gets. ☐

7 a. The more cramming you do the night before your exams, the better marks you will get. ☐

b. The important thing is to work 'smarter' not 'harder'. It's best to revise little but often, over a long period of time. It's a good idea to keep away from people who try to wind you up before the exam! ☐

Calming exam nerves

Professional sports people use this technique to prepare for big events. The more you can practise a situation in your mind (this is called 'visualization'), the easier it will get. You are giving your brain the positive signals that it needs, to cope effectively with a difficult situation.

Ask somebody to slowly read the sequence below to you, or record your own voice. First practise the 'Instant Calmer' (p. 21) till you are very familiar with it, and then tag this on the end. Make sure you are comfortable and won't be disturbed for at least 15 minutes.

You are feeling very relaxed and safe now.

It is easy for you to concentrate and see pictures clearly in your mind.

New thoughts help you change the things that you want to change.

These new thoughts go deep into your mind
and stay with you long after you open your eyes.

They allow you to cope better and better with life, day by day.

Now imagine you are lying on your own private golden beach.

Feel the sun gently warming your body. Feel a cool breeze on your face.

Listen to the sea, coming in and out, in and out, as you breathe in and out.

Breathe in calm, and breathe out any anxiety you have.

Smell the scent of the air. You feel safe and warm and happy and relaxed.

A little cloud appears in the blue sky, just watch it floating away.

If an unwanted thought pops into your mind at any time, just watch it float away without attaching to it. You are totally safe, feeling calm and chilled out.

In a few moments, you feel quite ready to leave the beach, full of energy and confidence . . .

You are going to take a walk in your mind along a little path, shaded by trees.

In the distance, you see an attractive looking building. What does the building look like? You enter the building, which feels calm and welcoming.

You look around and see a sign saying: 'Exam room'. If you feel any tension, let it go now. You breathe in calm, and breathe out any worries you have.

You stay feeling totally calm, and eager to see inside the room.

You open the door of the exam room and enter, feeling totally relaxed.

What does the room look like? Take a good look round. Look at the windows, the pictures on the walls, the floor, the furniture, the people in it . . .

A friendly-looking person comes over to you and welcomes you by name. Who is this person? Is it a man or a woman? Is this person somebody you know today or somebody who has passed on?

This person gives you something to bring you luck. What is it?

Now see yourself looking calm and confident as you take your seat.

You see an exam paper on the table before you. You calmly read the instructions and questions on it, feeling relieved. You just want to get started.

Enjoy that feeling of relief in your body. Enjoy the feeling of enthusiasm.

See yourself writing the answers as though you are watching yourself on a big cinema screen . . . You are coping so well. Your ideas are flowing.

Your memory is flowing. Your writing is flowing. You find the words you need.

Enjoy that feeling of flow and concentration.

Now you finish the paper and are ready to leave.

With each out breath, make the picture of the exam room get smaller and smaller. And now the building is getting smaller and smaller, as you walk away down the same little path you arrived by. Now the building is completely out of sight, like a pin head. All gone.

You find yourself back on your private beach.

In a moment, you feel quite ready to go home.

You will start to count backwards from 5 to 1.

When you hear the number 1, you begin to open your eyes.

You continue to feel relaxed, but energetic and well rested.

You feel ready to cope with any worries you have.

You feel confident and positive that you can give your best in any challenge that is thrown in your path. Enjoy that feeling of confidence in your body.

You know you can return to your own private beach anytime you want.

You feel calm and positive every time you think of the word 'exam'.

Now count backwards from 5 to 1.

Open your eyes slowly, move your body very slowly,

and come back to the room in your own time, feeling calm and rested.

You can adapt this visualization for any difficult situation, fear or weakness. It can really work. Use your own words and preferred images.

How to reduce stress – 10 top tips

Here are some strategies that students found worked for them:

- **Don't wait for anybody to make life better for you.** Take control and take action – despite obstacles or negative past experiences.
- **Choose a role model carefully** and **don't put off talking to someone you trust,** when you need to.
- **Be kind to yourself and others** – especially if anything goes wrong. Learn to accept that no human is ever perfect. No situation is ever totally black or totally white. Instead ask yourself 'what can I learn from this?' You can't always change a situation, but **you can change your attitude** to it – then your feelings will also change.
- **You don't always have to say 'yes'**, when asked to do something you really don't want to do. Practise saying: 'I choose to' instead of 'I have to', and feel the difference ...
- **Use your time and energy wisely**. Simplify your life as much as possible; **delegate** to others what you can't do yourself. **Get rid of unrealistic ambitions,** but don't discard hope, as you **work with persistence and enthusiasm towards your aims**.
- **Set aside some time to be alone to review** your current stress levels and evaluate your activities and goals.
- **Look after your body and mind.** Walk as much as possible – especially in nature. Don't treat your body like a rubbish bin. Drink lots of water.
- Use every free opportunity to **relax and do something you love**: laughing releases anti-stress chemicals into your bloodstream. Practise a relaxation and breathing technique regularly, so that you can use it automatically in an emergency or before an exam.
- **Let go of outdated beliefs**: Be aware that you may be carrying around old negative messages that you received as a child, and still be behaving as though they were true. We are all resistant to changing these messages, as we sometimes prefer to hold onto a familiar idea of ourselves rather than take a risk of changing.

  If you get caught up in unwanted thoughts or feelings, be aware that they are just temporary – thoughts are not reality, they just come and go in the mind. Imagine each thought like a bubble floating away down a stream ...
- **Reframe unhelpful thoughts**: Turn unhelpful thoughts into positive ones, for example:

  'Everyone seems to learn things faster than I do ...'
  turns into:
  'It doesn't matter about the others. I'm doing my best.'

  'I'm absolutely useless at exams, I know I'm going to panic.'
  turns into:
  'There's nothing I can do about the past. But I do have control over the next 10 minutes. I'm going to practise some rhythmic breathing.'

Chapter 2

Bodymaps

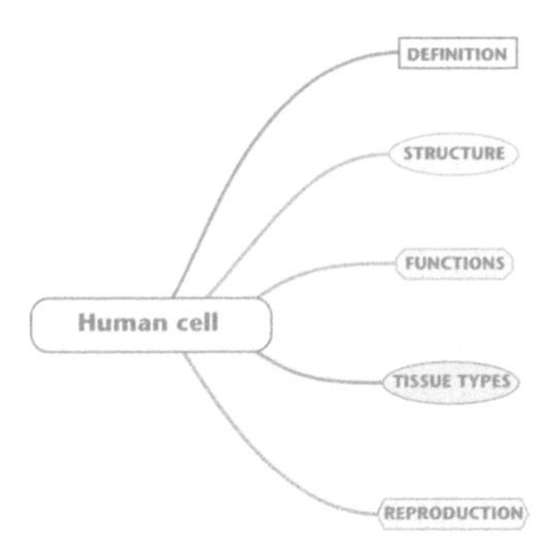

OVERVIEW

This chapter offers you opportunities to explore what a Bodymap is and how to make one.

You will be able to practise the different uses of a Bodymap, including:

- ✓ **for exam revision;**
- ✓ **for boosting motivation;**
- ✓ **as a planning tool.**

What *is* a Bodymap?

A Bodymap is a system to help you express, organize and remember information in a *visual* way. Making Bodymaps is one of the best ways to help you memorize anatomy and physiology. To get a clear idea, take a quick look at the Bodymap of 'the Human Cell' on page 30.

What does a Bodymap usually include?

- **key words** (as few as possible to express ideas);
- **abbreviations** (shortened forms of words);
- **lines** (that connect ideas together);
- **arrows** (that show relationships between ideas);
- **pictures and symbols** (that illustrate ideas);
- **colours** (to organize or highlight ideas);
- **numbers** (to put ideas in order).

Why are Bodymaps unique ?

All Bodymaps have a similar structure. But each Bodymap will look quite different from the next one. For example, trees usually have a trunk, branches and leaves, but there is no end to the different kinds of tree that you will see.

Because each Bodymap is unique, it will stand out in your brain and instantly be easier to remember. As you know, endless pages of boring lines of printed words only put your brain to sleep, when it is begging to be stimulated.

Why do Bodymaps work so well?

Bodymaps work well because they make use of more of your brain cells than usual. They allow you to see both the whole picture of a particular topic, as well as the smaller parts that make it up (like a jigsaw puzzle).

How do you make a Bodymap?

Making a Bodymap becomes easy with a little practice. One kind of Bodymap that you may find useful as a starting point is based on the idea of the 'Mind Map'. Tony Buzan invented Mind Maps to help people make better use of their brains. He said that we all use just a tiny little part of our brain potential. The best way to understand a Bodymap is to make one yourself. Just follow the **6-step system** below until you develop your own personal style . . .

The 6-Step System

Step 1: Equip yourself

- You will need at least 2 sheets of paper: A3 paper is a good size to start with. You will use at least one sheet for rough work and another to copy out a final Bodymap.
- Use the paper in landscape position:

This way round . . .

Not this way up!

- You will also need lots of different coloured felt tips, but some coloured pens or pencils will do.

Step 2: Brainstorming

- On your first piece of paper: write the topic you want to tackle in large capital letters in the middle of the page. (You can also draw an image or symbol to represent that idea.)
- Then draw a circle round it, so that the idea is inside a bubble.
- Round this bubble, quickly jot down any ideas about the topic you are tackling. (See below for an example.)

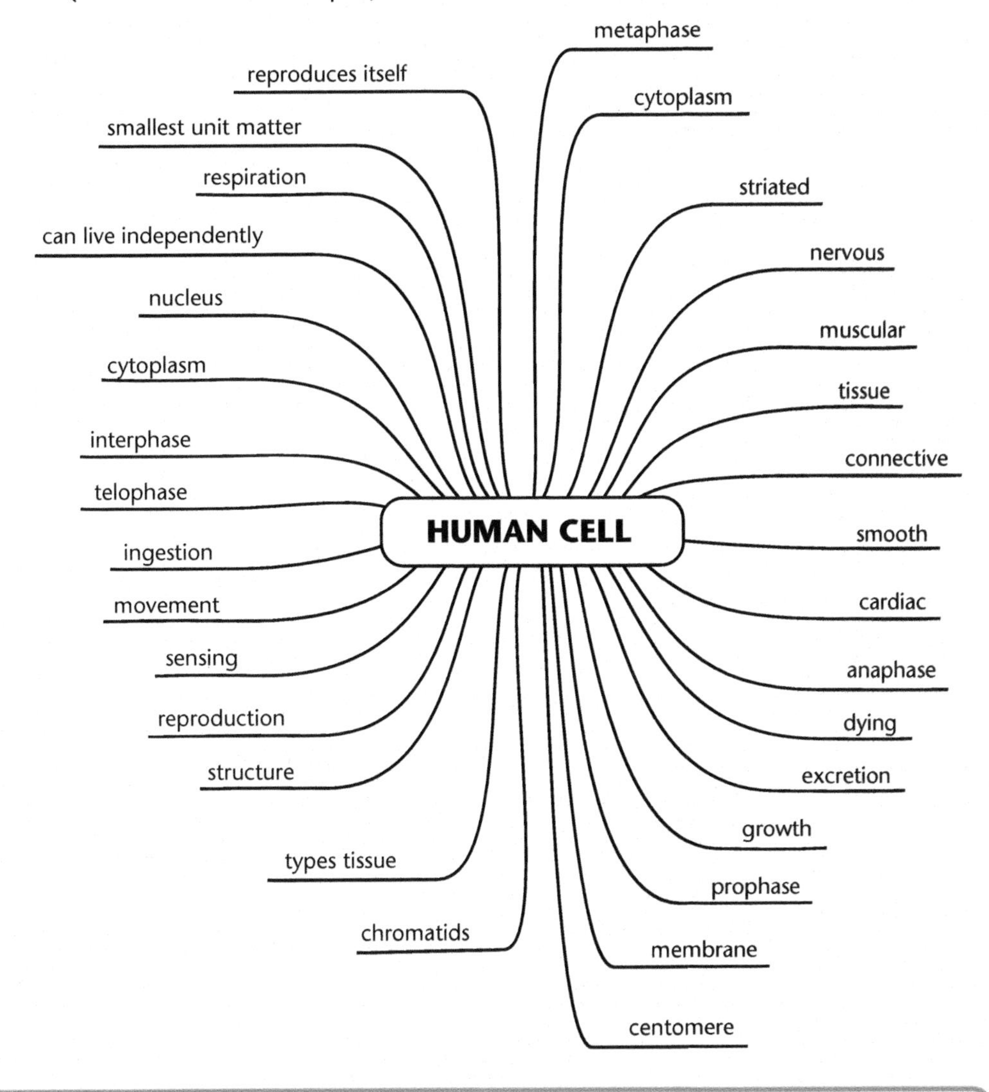

This step is called 'brainstorming' because you just let your brain come up with as many ideas as possible, as quickly as possible, in any order you like. Don't worry about spelling or putting anything 'wrong' down. You can change or get rid of ideas later.

Step 3: Colour ideas that go together

- Find which ideas seem to go best together.
- Colour all the ideas that go together in the same colour (this is called 'colour coding'). Use highlighter pens, felt tips or pencils.
- If any idea doesn't seem to fit or isn't useful, just cross it out.
  Look at the example below: our main topic is the Human Cell:
  - Everything to do with **functions** of the cell is coloured **blue**.
  - Everything to do with **reproduction** for the journey is in **red**.
  - Everything to do with **structure** is in **green**.
  - Everything to do with **tissue types** is in **brown**.
  - Everything to do with the **definition** is in **black**.

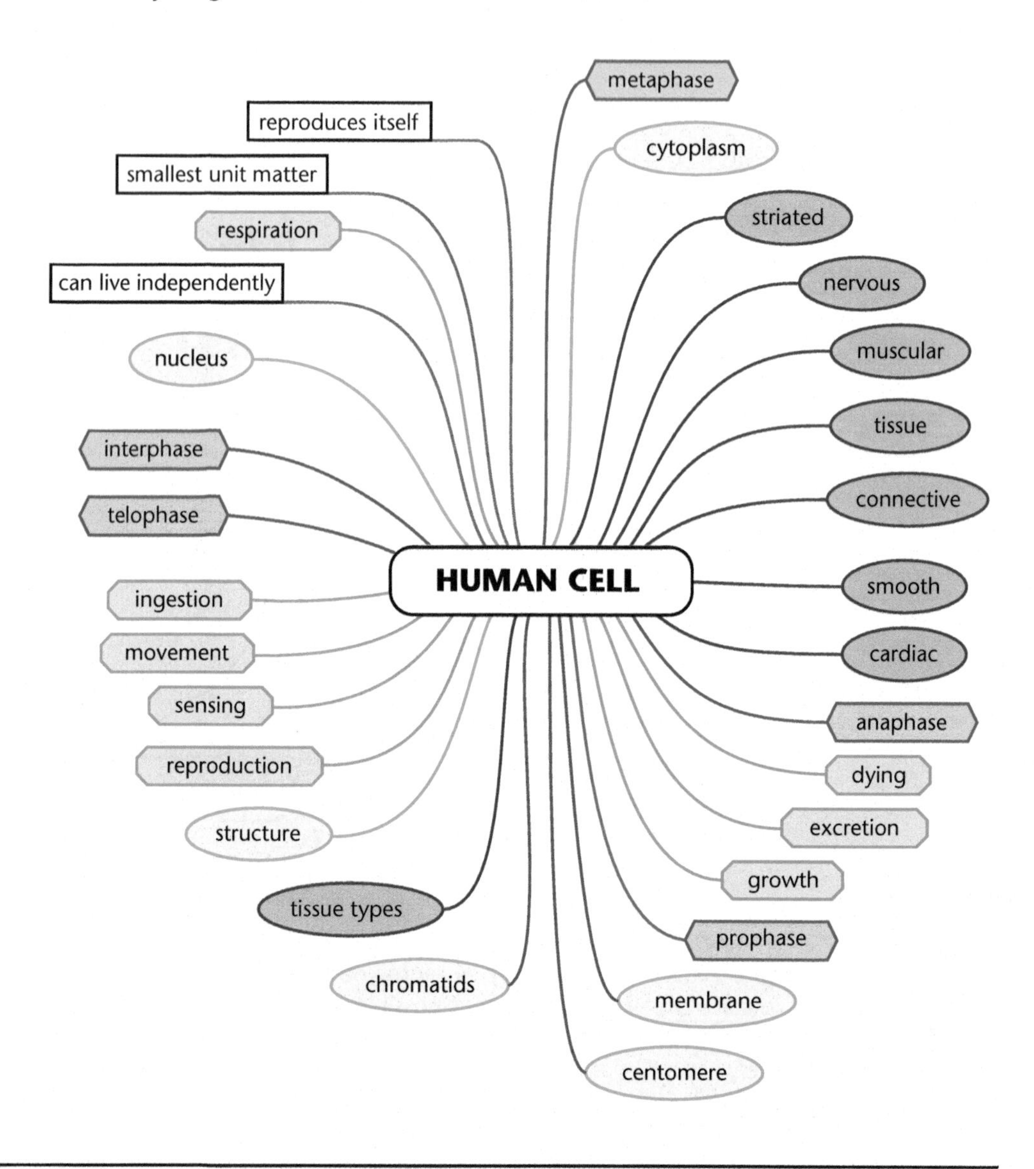

Step 4: Deciding about headings for main ideas

- Now take a fresh piece of paper.
- Once again, put the title of your topic in the middle of the page, inside a bubble. Add an image or symbol if you like.
- Find the best **key** word that describes each **main idea**. Now write a key word for each main idea that branches out from the central bubble. Keep your writing as horizontal as possible.

Use a coloured pen to draw a thick supporting line from your main topic to your key words. Use a different coloured pen for each main idea. Some people draw a horizontal line under the key word to support it like a shelf. Other people prefer to draw a bubble or box round each main idea. See below for an example.

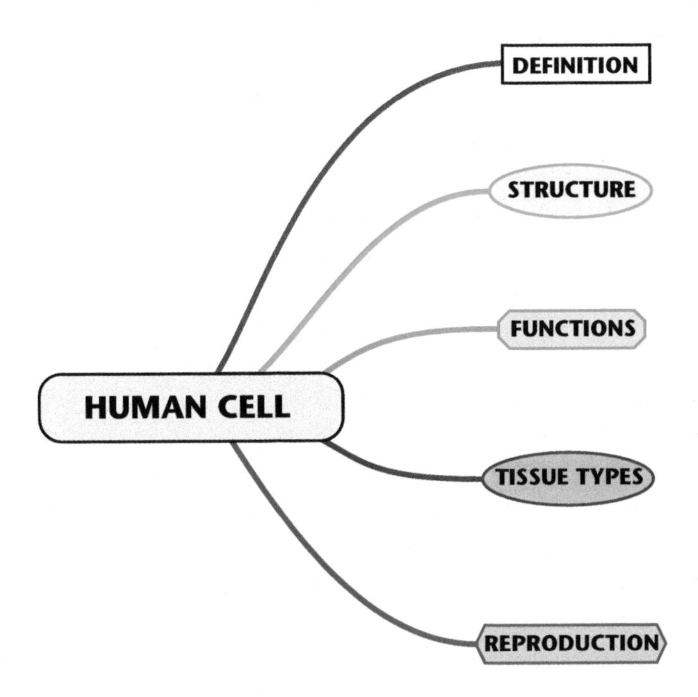

Step 5: Organizing and adding more information

- From each thick main branch, draw **smaller branches** for **each related idea** (see page 33).
- You can go on drawing smaller and smaller 'twigs', depending on the level of detail you want to explore. But it isn't helpful to make your page too crowded.

Step 6: Finishing touches

The finishing touches you use are very important, as they will give each Bodymap its unique appearance (see page 34). You may want to add the following:

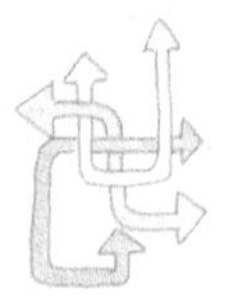

arrows to show how an idea relates to another;

symbols and pictures to illustrate points and make them more memorable;

an enclosing line round clusters of related ideas;

numbers to sequence your points.

What if my Bodymap looks *terrible* ?

Please don't worry if your first attempts are messy or disorganized: the more you re-think and re-draw your Bodymap, the more you will **fix memories in your brain**. There are some ready-made Bodymaps in this book that you can use, but it is usually best to make your own. The physical and mental act of making your own unique Bodymap will be the most powerful memory aid you will find. *'But I'm useless at drawing . . . I don't have any imagination!'*, I hear some of you insisting. Drawing skills are *not* important, and you will find that, as you get more and more used to the Bodymap system, you will start *thinking* more flexibly and imaginatively.

There is excellent computer software for making Mind Maps (see page 171 for details), but it's a good idea to start off making your Bodymap by hand and then use software to easily create some stunning Bodymaps.

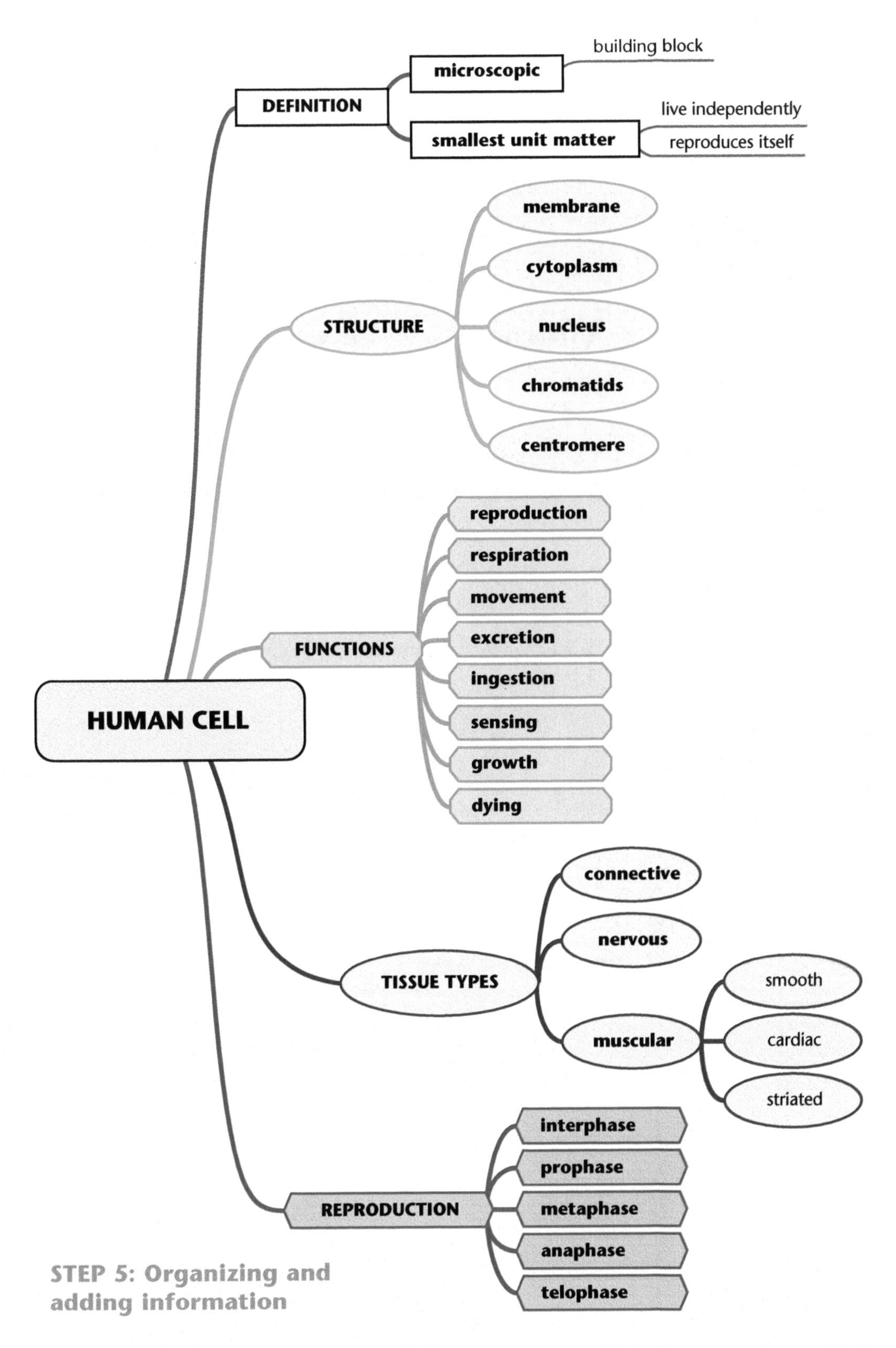

STEP 5: Organizing and adding information

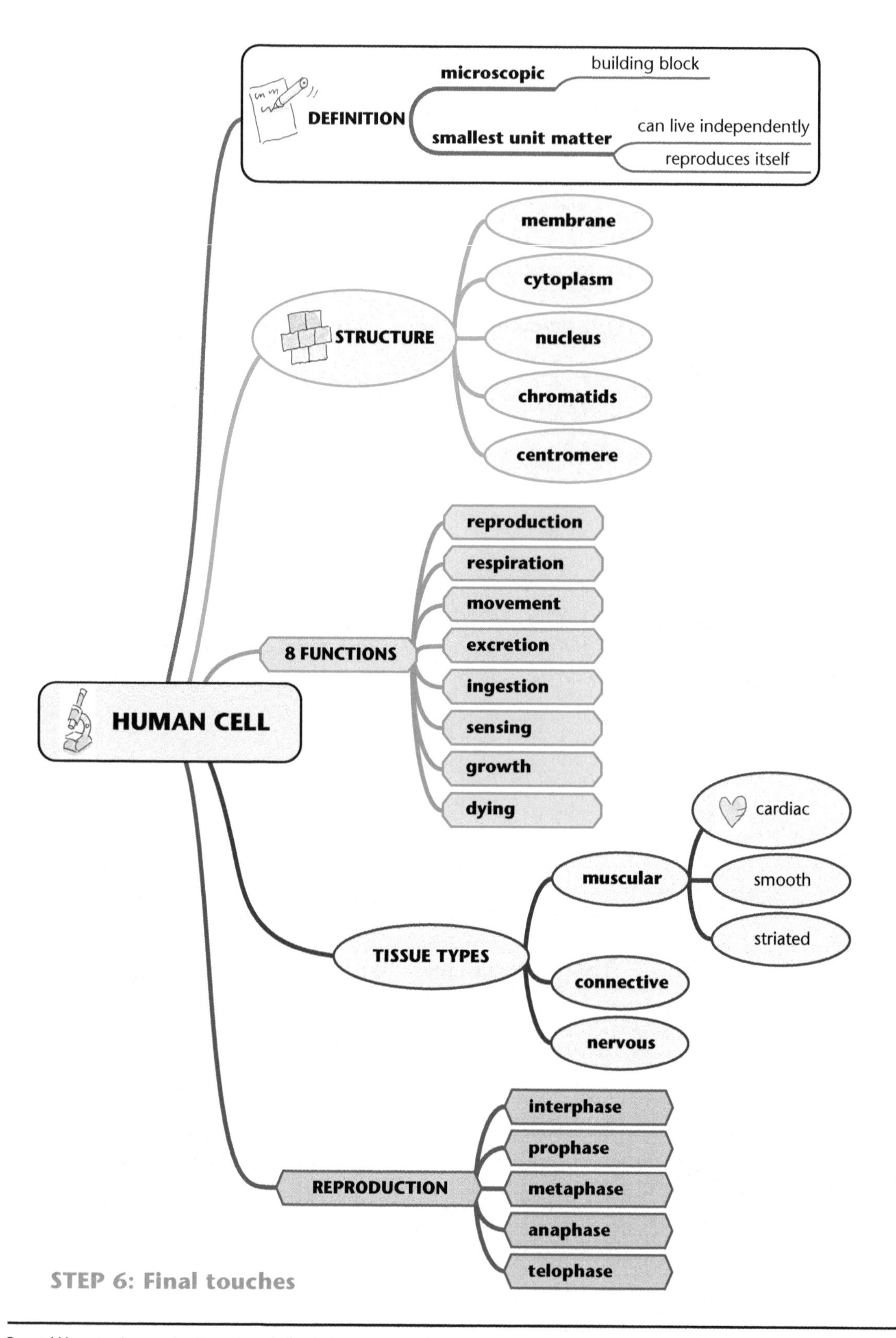

STEP 6: Final touches

Key words

You will need to make really good notes:

- during lectures;
- when reading;
- for exam revision.

Your notes will be useless if

- they are too long and wordy;
- you have missed out essential points;
- you can't understand them when you read them back.

So it is really important to only write down **key words** when you are making notes.

Key words are those words that are the most important to express an idea: they unlock the meaning of what you want to say. If you took all the key words out of a sentence, it would mean nothing at all! If you can't express your idea in one key word, you can use up to three, but no more.

Example:

'Remember to disinfect the wound.'
disinfect and **wound** are key words

'Digestion of food particles'
Digestion is the key word

Bodymap: summary

- Let your imagination run wild.
- Look for links between ideas.
- Use images.
- Use key words only: no more than three per branch.
- Don't worry about making your Bodymap look good at first.

What else can I use a Bodymap for?

Apart from using Bodymaps to boost your memory, you can use them for:

- taking and presenting course notes;
- planning an essay or examination answer;
- problem-solving;
- group work;
- time management;
- finding ideas;
- making a spoken presentation; and
- exam revision.

How can Bodymaps be used for exam revision?

Bodymaps make A&P easier to remember

Page after page full of boring print can send the brain to sleep – just like driving for miles along a straight, dull motorway. With Bodymaps, the information can look much more exciting, so the brain stays switched on.

If you need to remember all the information on your Bodymap, the '**Re-draw and Compare Method**' *really* works:

1. **Look** over your Bodymap really carefully for a few minutes. Then **Cover it up**.
2. **Redraw** your Bodymap from memory on a new page.
   Don't worry about colour, illustrations or spelling.
3. **Compare** both Bodymaps, noting carefully the parts that you have missed out.
4. **Re-draw** your Bodymap once again.
5. Compare again with the original Bodymap, noting carefully what you have missed out.
6. When your **latest Bodymap is similar to the original**, you will know that you have created a strong memory – but probably not a permanent one as yet.
7. To **fix your Bodymap into your long-term memory**, it is important you revise it at **particular points in time**.
   You will need to **repeat** steps 1 to 6:
   - on the same day
   - 24 hours later
   - 3 days later
   - week later
   - and 1 month later

If you don't revise within 24 hours, you may forget nearly all the information that you have been trying to remember. If you have a lot of problems with memory tasks, you may have to revise even more often.

Make yourself a large wall poster of important Bodymaps. An A3 poster should be a minimum size, but the bigger the better. Some students use cheap rolls of wallpaper lining paper to make giant maps.

Using a Bodymap as a planning tool

Use this map to tick off what you learn as you progress in your studies.

You can also use it to plan your revision programme.

You may want to create your own giant revision map with more detail, colour and symbols (for example by adding symbols such as the heart on the cardio-vascular system).

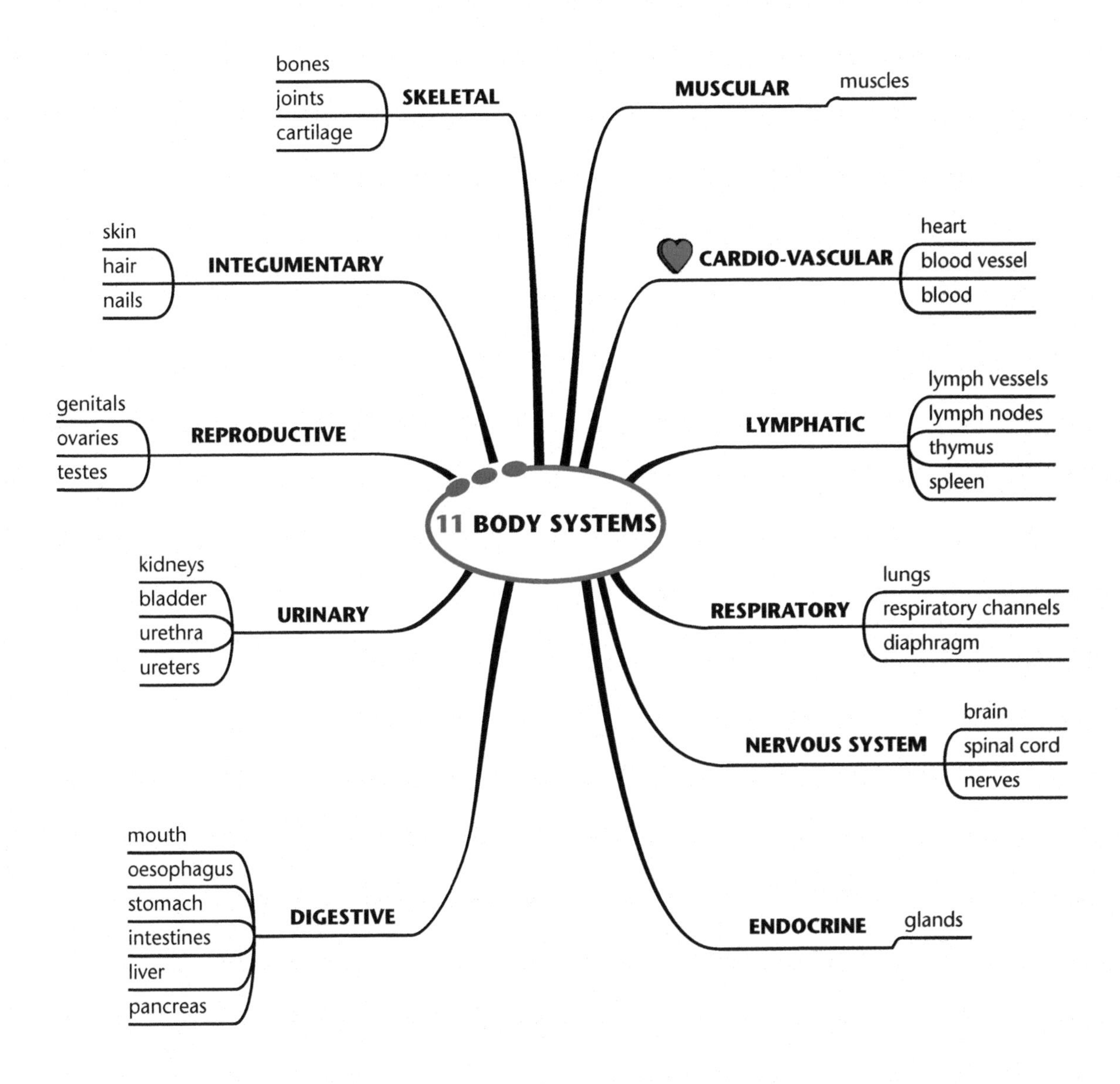

Chapter 3

Timing is everything

OVERVIEW

This chapter presents strategies to help you work smarter not harder. If you manage your time well, you might be able to study effectively and have a life! You will have opportunities to:

- ✓ **take a good look at how you are spending your time at the moment;**
- ✓ **try out some systems to help you plan your time better and save energy.**

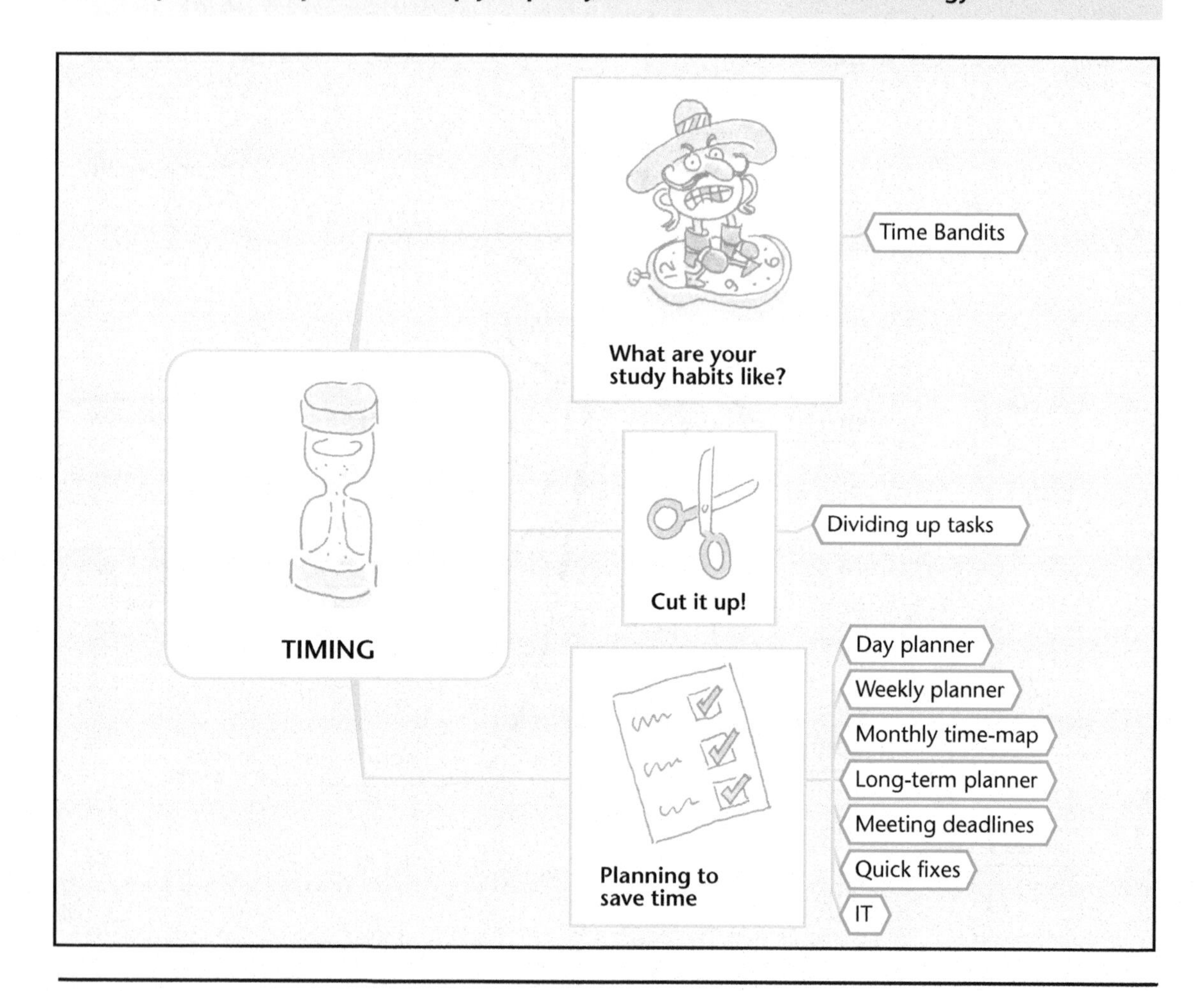

Questionnaire: What are your study habits like?

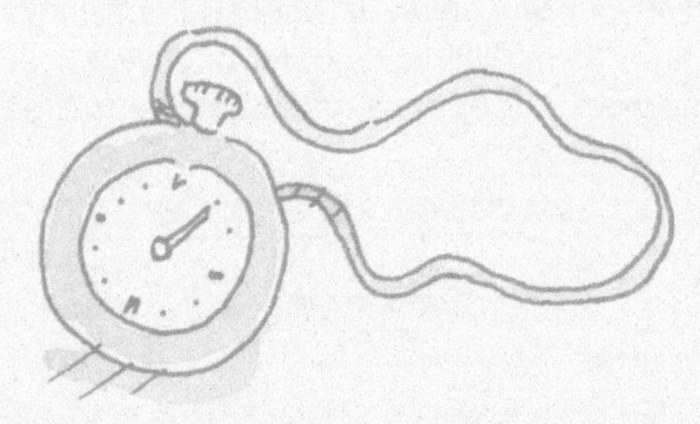

Which of the statements below sounds like you?

1 I put off tasks that I don't like for as long as possible. ☐

2 I find it hard to concentrate for long on things that don't interest me. ☐

3 I usually flit from one task to another (such as phone, email, reading, social media, TV, chat, gaming etc.). ☐

4 I'm often too shattered to study, especially if I've had a late night. ☐

5 I've got no weekly or monthly plan for my studies. ☐

6 I don't recopy or file my notes after lectures. ☐

7 My notes are a mess and I usually can't find what I want. ☐

8 I do great bursts of work one day and nothing the next. ☐

9 I get behind with my revision and cram at the last moment. ☐

10 I don't have any special memory techniques. ☐

11 I don't use a list, calendar or diary to prepare what I need for the next day. ☐

12 I don't know where and when I work best. ☐

If you have answered 'Yes' to any of the above, it's time to get *organized*. See the following pages for solutions that have worked for other students.

Know your Time Bandits

Time Bandits steal your time away without you noticing.
To keep a watch on them, fill in a diary for one week.
At the end of the week, tot up how much time your Time Bandits have taken.
Did you get any surprises?

Time Bandit	Time spent
Personal study	
Attending course	
Paid work	
Household tasks	
Sleeping and eating	
Exercise	
Chilling out	
TV	
Travel	
Going out	
Anything else . . .	

Now, using the table below, decide on an ideal amount of time to allow for each of your weekly activities. Make a list of suggestions how to do this: work with somebody else if possible.

Time Bandit	Ideal time to spend on this:
Personal study	
Attending course	
Paid work	
Household tasks	
Sleeping and eating	
Exercise	
Chilling out	
TV	
Travel	
Going out	
Anything else . . .	

At the end of the month, fill in the Time Bandit chart again and see whether you have managed your time better.

Cut it up!

Cut a big memory task into bite-size chunks

You are wasting your efforts if you try to learn too much information all at the same time. As you pour more into your brain, it just leaks out again. A 20-minute burst of study, followed by a short break and then a self-test to refresh information, is *far* more effective than a longer study period.

Start by **cutting a big memory task into bite-size chunks.** Forget all other chunks while you are focusing on a particular one. Thinking of how enormous and impossible the task seems, will leave you feeling hopeless before you even start.

For example, when attempting to learn the bones of the skeleton, divide the task into 8 chunks:

- the skull
- the vertebrae
- the torso
- the arms
- the hands
- the sacral area
- the legs
- the feet

See page 43

Half-learnt information will interfere with the learning of new things

For example, *don't* go on to learn the vertebrae until you can draw and label the skull with ease. *Don't* compare yourself with other people's rate of learning: just follow your own rhythm and capacity. You've heard of the tortoise and the hare, haven't you? The tortoise went slowly and surely, and got a distinction in its A&P exam.

See page 44

Memorizing similar information sets up confusion

For example, keep the foot and the hand apart! Do not learn the bones of the feet and the hands at close intervals to each other: you're sure to mix up your metatarsals with your metacarpals, as well as confusing the small bones.

See page 45

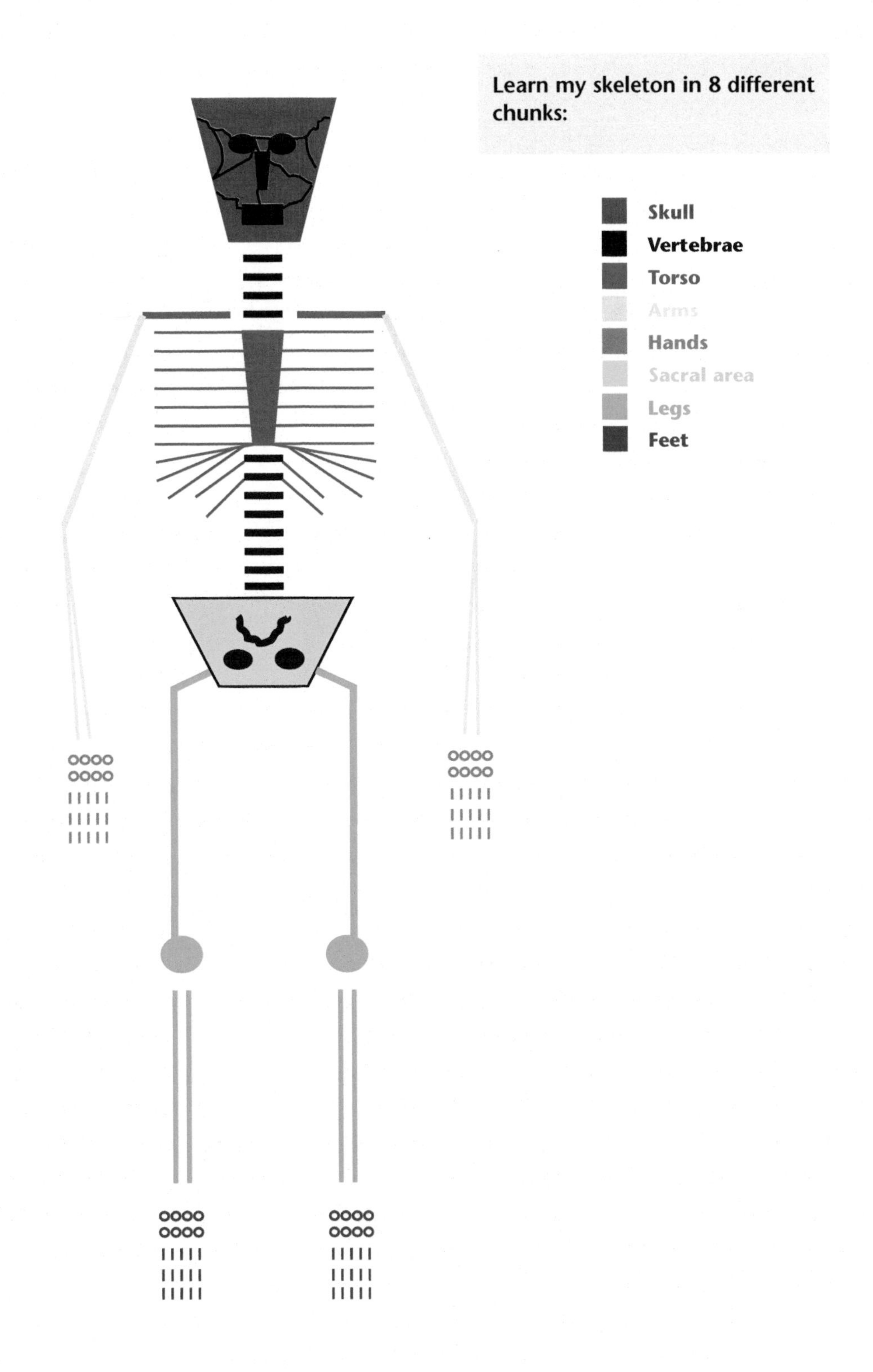
Learn my skeleton in 8 different chunks:
Skull
Vertebrae
Torso
Arms
Hands
Sacral area
Legs
Feet

Learning the skull

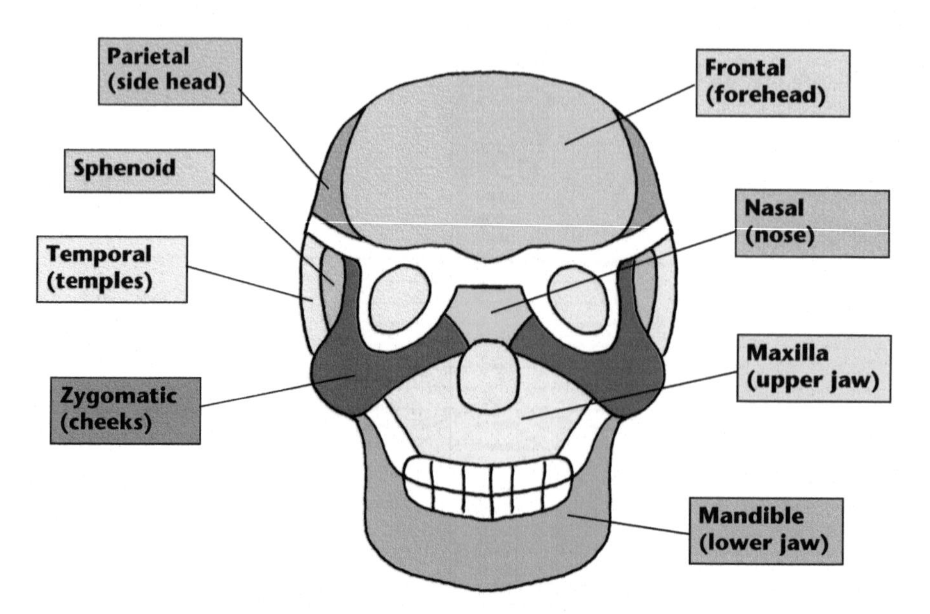

Learning the skull

1. **Simplify**: make a very simple drawing of the skull – this can be developed at a later stage.
2. **Prioritize**: first learn only the eight most important bones.
3. **Make the picture stand out in your mind**: use a different colour for each bit (red for cheeks, brown for temple etc.). I drew glasses round the eyes: it helped me remember the position of the bones.
4. **Divide the task into smaller tasks**: learn the bones on the right side of the drawing first, then the left side.
5. **Use all your senses**: touch the different bones in your face as you say them in order from the top down: frontal, nasal, maxilla, mandible . . .
6. **Make up memory tricks**: I called the upper and lower jaw: 'Maxi' and 'Mandi' – silly, but it works!
7. **Use the 'WAM' spelling method**: it's hard to remember words that are difficult to say and spell, like 'sphenoid'. (See Chapter 8.)

Learning the hand and foot bones

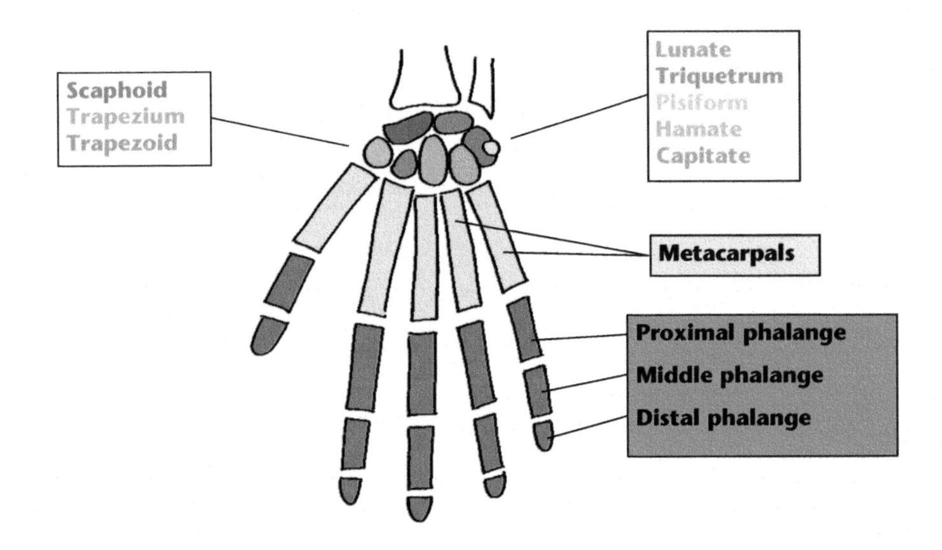

Keep the hand and the foot as far apart from each other as possible!

Do not start to learn the foot until you know the hand perfectly . . .

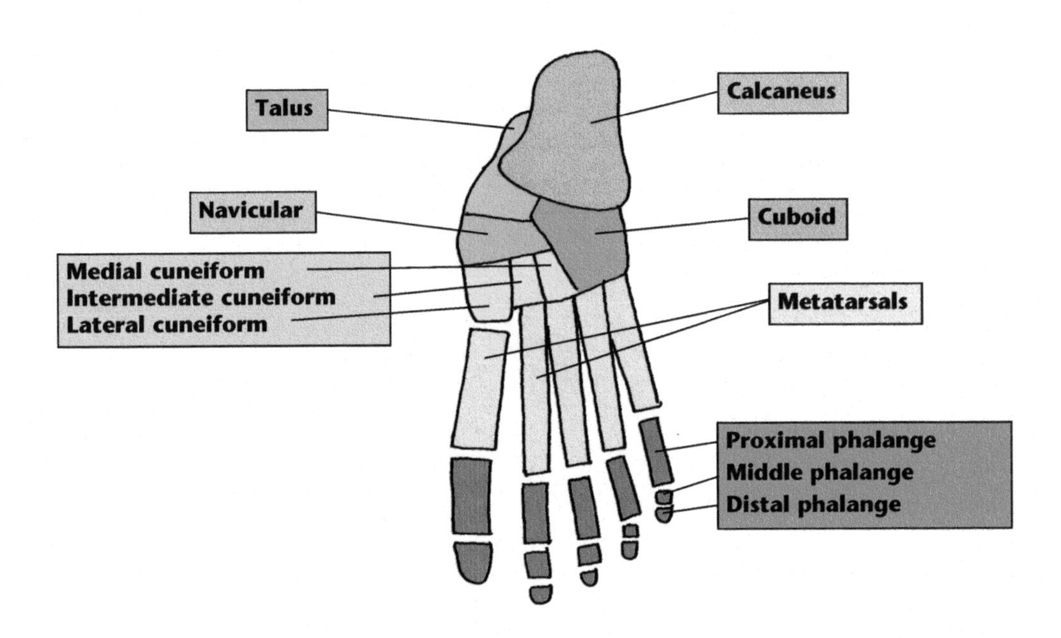

Activity: Muscles of the lower arm

When you learn the **muscles of the lower arm** learn them in 2 separate groups:

- Learn the muscles in **Group 1**.
- Learn the muscles in **Group 2** *only when you are totally confident about the first group.*

Group 1	
1. **Brachialis**	flexes **elbow**
2. **Brachioradialis**	flexes **elbow**
3. **Pronator Teres**	flexes **elbow**

Group 2	
4. **Flexor Carpi Radialis**	flexes **wrist**
5. **Flexor Carpi Ulnaris**	flexes **wrist**
6. **Flexor Digitorum**	flexes **fingers**

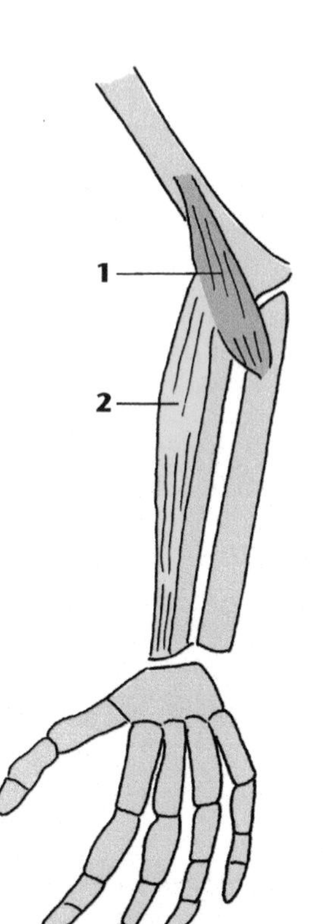

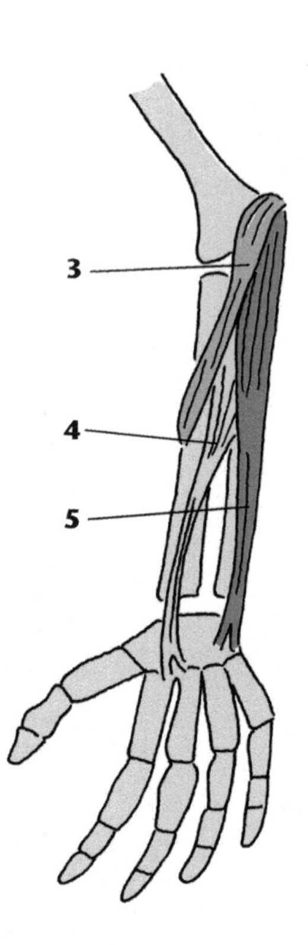

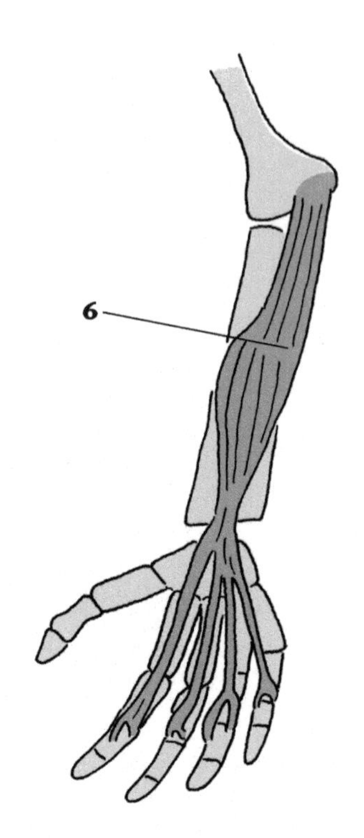

Saving time

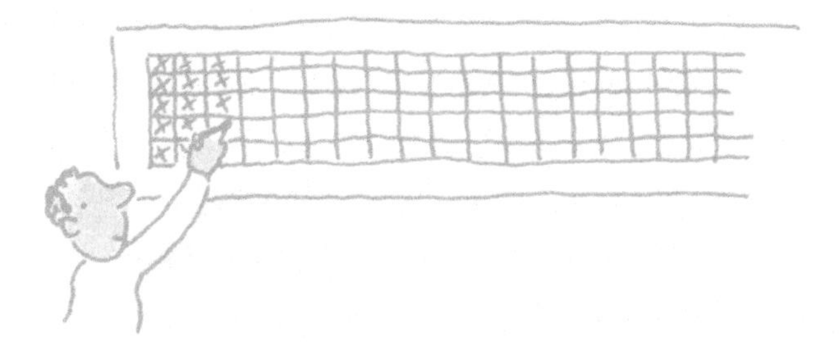

Make a long-term timetable (see page 52 for Long-term Planner)

- Write deadlines for assignments, exam dates and appointments on a big wall chart or use the Long-term Planner. Also give yourself a week's warning before important events.
- Make yourself an exam revision programme and mark revision tasks on your Planners. It is essential to actively memorize each item you want to remember a number of times and at particular intervals in order to lay down strong memory traces in your brain. (See page 51.)

Make yourself a monthly time-map (see example on page 51)

- Use the Bodymap technique. Mark what is '**urgent**'; what needs to be done '**soon**'; what is '**awaiting**' some action from somebody else; what is an '**ongoing**' **activity;** and what you want to do '**whenever**'. Prioritize tasks according to what must get done, not what you like best!

Make a weekly timetable (use the Weekly Planner on page 50)

- Balance commitments with relaxation and try to stick to your plan.
- Share tasks with a friend: ideas, lecture notes, research, revision, proofreading.
- Vary activities: writing, reading, filing, revision, recopying notes etc.

Daily routines (use the Day Planner on page 49)

- Break larger tasks into small steps and set yourself a realistic timed task. Use a brainstorm technique if you can't get started.
- Work in short bursts of 20 minutes, followed by revision, a short break, then try to recall what you were doing previously.

Note-making after lectures

- Rework notes and recordings as soon as possible after lectures.
- Use highlighter pens to group ideas that go together. Write each major point on a separate page with a heading, or use a Bodymap.
- Don't write in full sentences: use headings, key words, abbreviations or pictures.
- Label your notes with a title and date. File as soon as finished.
- Alternatively, record your notes onto a recording device.

Reading

- Read only what is *relevant* to your task in hand. Ask a librarian or lecturer to help you make the right choices: sometimes you may be able to find an 'easier' book, internet site or video with the same information in a better format for you.
- Be clear about *exactly what you want to know before you start reading.* Make a copy of important bits of text and underline relevant points with a pencil or highlighter.
- You may be able to scan in useful bits of information into a PC and then use read-back software to help you listen to the text as you read. Ask your librarian for advice.
- If you will have to write a bibliography at the end of an essay, for each text you read, complete an index card with:
  - the title and author's full name;
  - the edition and place/date of publication;
  - reference number;
  - key ideas, quotations and page references.

  This will save you a lot of time later on!
- Keep exact details of information that you find from the internet. Remember, your work will not be valid if you try to pass off somebody else's writing or ideas as your own.

See Chapter 9, 'Read right', to improve your reading speed.

Day Planner

WHAT I HAVE TO DO TODAY – Date:

	Urgent tasks	✔	People to see People to phone (+ tel. number)	✔
1				
2				
3				
4				
5				

Notes (including developments that are awaited and items needing further attention)

Weekly Planner

Date	Morning	Afternoon	Evening
Monday			
Tuesday			
Wednesday			
Thursday			
Friday			
Saturday			
Sunday			

Monthly Time-map

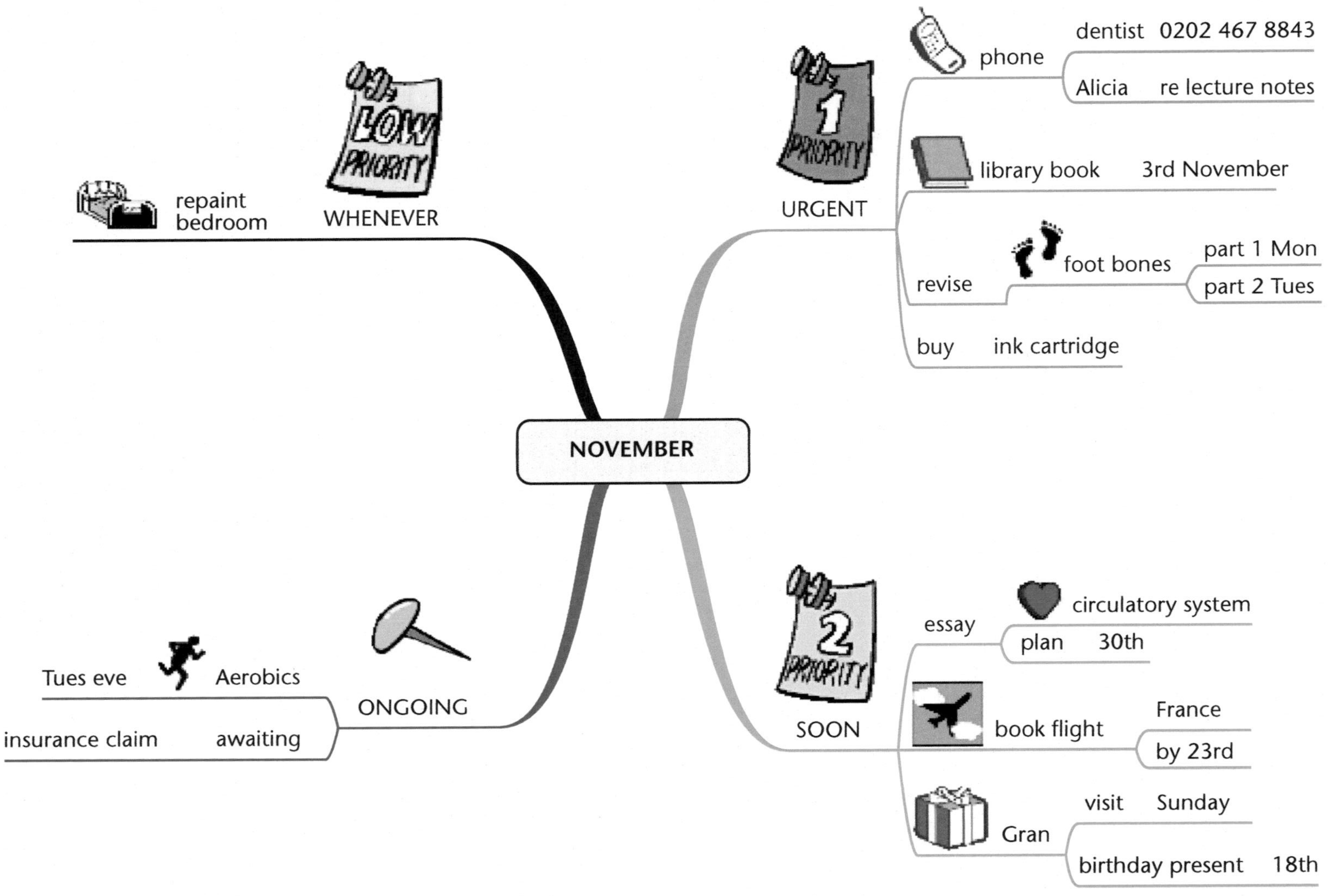

NOVEMBER

WHENEVER
repaint bedroom

URGENT
phone
dentist 0202 467 8843
Alicia re lecture notes
library book 3rd November
revise foot bones
part 1 Mon
part 2 Tues
buy ink cartridge

ONGOING
Aerobics Tues eve
insurance claim awaiting

SOON
essay
circulatory system
plan 30th
book flight
France
by 23rd
Gran
visit Sunday
birthday present 18th

Long-term Planner

Day	Month	Month	Month
1			
2			
3			
4			
5			
6			
7			
8			
9			
10			
11			
12			
13			
14			
15			
16			
17			
18			
19			
20			
21			
22			
23			
24			
25			
26			
27			
28			
29			
30			
31			

You can use this for planning ahead over a 3-month period. Note down dates for: starting and finishing assignments; different exam topics for revision; and any other important events.

Meeting assignment deadlines

Step 1: Work out how many days you think it will take you to:

a. **Get started**. This will include:

- working out what exactly you are being asked to do;
- brainstorming and thinking through your ideas; talking about your ideas with others;
- and doing a rough plan.

b. **Do your research**. This will include:

- working out what information you need;
- collecting relevant information.

c. **Organize your information**. This will include:

- selecting exactly what to include and what to leave out;
- grouping material and organizing it into a logical order.

d. **Make drafts and rough copies**. This will include:

- writing up your ideas in a first rough draft;
- carefully re-reading your writing, looking for ways to improve it – you may need to ask somebody else to read over your work and then you may have to re-write some bits;
- correcting any spelling or grammar mistakes.

e. **Complete the task**. This will include:

- putting together the bibliography;
- producing a final fair copy ready to hand in.

Step 2: Add up the number of days you think Step 1 will take you.

Add a few extra days 'just in case . . .'

Step 3: Take that number away from your deadline date.

This will give you your starting date. Mark this on your Long-term Planner or calendar.

Step 4: Keep a running total as you work.

You may have to readjust the time you allocate to each task. Remember that you cannot push back the deadline.

Quick fixes for time-saving

Here are some tricks that students found worked for them

Time is money: 'I've worked out that on a salary of £20,000, one minute of my time costs 20p and ten minutes cost £2. If I waste an hour, I might as well burn a £10 note and throw two pound coins in the bin! Every day I look at my weekly plan and make a list of what's most important to do that day – and usually I stick to it.'

'I do the **Twenty-minute burst**: I set the alarm on my watch and work for 20 minutes, then I have a 5 minute break. Start off again with a quick revision – that bit is essential. It's apparently *the* best way to learn something.'

'Every day I go to my desk with a bin bag and ask myself: "**What will happen if** I throw this away?" This is the only way I can clear my desk. Then I start work: I switch off my mobile and put a "Danger, Keep Out" sign on my door. I wear earplugs or headphones at college, so I can't hear any background noise.'

Russian Roulette: 'I list all the tasks I hate on little pieces of paper and put them in a bag. Then I pick one out without looking – that's the one I've got to start with. It's better when you do it with your pals, you can share out tasks like that!'

Bribery: 'I choose a task I don't like and divide it into small achievable steps. I tackle the first step when I am feeling at my best, and I bribe myself. I have a jar full of 20p coins. Every time I do an hour's work, I put 20p in. At the end of the month, I go mad and spend the lot.'

Imagine: 'When I have a boring task to do, I remind myself why I am studying. I close my eyes and imagine in great detail what it will be like when I have passed my exams. I then make a list of all the advantages of getting the task done. Next, I write down all the disadvantages of not getting that task done.'

'I did things my way: I started revision for exams in week 1 of my course! When it came to the exams, I was really cool. I redid my notes after each lecture, and stuck to a big revision chart I have on the wall. I didn't use the set books. I've found some CD ROMs and internet sites that suit me better.'

Resist the lure of the screen: 'I was getting totally addicted to being connected 24/7, spending longer and longer each day checking emails, tweeting, gaming, downloading music. I totally resist turning on my phone when working with other students or during study. I turn it off when walking, eating and talking with people!'

IT solutions for saving time

There are many technological aids to help you save time when studying. Additionally, there are some specialized tools for studying anatomy and physiology. You can use these to:

- read and spell medical terms;
- remember and pronounce medical terms correctly;
- make notes and complete written work;
- visualize body parts and functions;
- record and edit spoken information;
- revise for exams.

Medical spellcheckers, dictionaries and word lists

Handheld spellcheckers and spellchecking software: these may cover about 150,000 medical terms. Some include a UK or American dictionary, speech function, calculator, databank and anatomical illustrations. Some cover more than one topic, including dentistry and pharmacy.

Word lists: you can buy writing tools that sit alongside your word processor, giving you rapid access to lists of words that you frequently need. You can enter your own personal lists quickly and easily.

Visual Thesaurus: you can buy software to explore the meaning of words. When looking up a word, it creates an animated webbed diagram with your word in the centre of the display, connected to related words and meanings. It includes a spell checker, word suggestion and text-to-speech.

Aids to reading

Scanning pens: these pens can be used to scan in a medical term from a book or handout, hear it read back and see a definition. American dictionaries recognize many UK terms.

Turning text into speech: you can use text-to-speech software to have text read aloud to you. You can also use it for proofreading your own writing. Some of these programs also contain a phonetic spell checker, word predictor, dictionary, research tools or speaking scientific calculator.

Aids to note-taking

Recording devices: you can use small digital machines to record your lectures, ideas or discussions. You can get **software, such as Audio Notetaker**, to help you review and organize key sections of your recordings. This makes the task of re-listening to long digital recordings quicker and easier. You can also listen to recordings of your documents while on the move. You can write out your notes at a later point, if you wish.

Aids to writing

Speech Recognition: medical editions of speech recognition software packages allow you to create written documents, notes and emails just by speaking into your computer. When used with a digital recorder, you can make spoken notes wherever you are working and transcribe them later. You will need a quiet environment to work in and time to train the software to recognise your voice.

Mind Mapping software

These software packages are flexible tools for brainstorming, planning, noting, organizing, reviewing, memorizing and presenting information in map or chart format. Essential for creating dynamic bodymaps!

Memory tools

There are websites that provide free tools to help you learn any subject, including a vast database of flashcards. They include facilities for tracking your progress and testing yourself, as well as game modes to help you enjoy memorizing your subject.

Visualizing the human body

This kind of software provides memorable multi-sensory animations, illustrations and 3D models of inside the human body.

It's very important to get good advice about the best product to meet your own particular study needs. Also, some initial training can save you a lot of frustration and help you use your IT tools efficiently. Technology changes so quickly that information about products is almost out of date as soon as it appears. Common options are explored above, but always check with suppliers for the latest solutions. Some favourite student recommendations can be found in the 'Useful Resources' section on pages 169 to 171.

Chapter 4

A picture is worth a 1000 words . . .

OVERVIEW

This chapter will help you understand and explore the full potential of visual learning. It offers you opportunities to practise strategies for visual learning such as diagrams, cartoons, Bodymaps, lists with bullet points, colour-coding, flash cards, posters and bookmarks.

Visual learning

The old saying 'A picture is worth a thousand words' is backed up by solid research on how we remember information most effectively. We store information in the brain in both words and images. That is why we can learn almost twice as well from words and images as from words alone. Visual learning engages both hemispheres of the brain and is best suited for anything that involves accurate recall of information.

Visual learning techniques help you to:

- **Simplify complex abstract information**
  You can use visuals to display large amounts of information in ways that are easy to understand.
- **Organize and analyse information**
  You can discover how ideas are connected and how information can be grouped and organized. New concepts are more easily understood when they are linked to something you already know.
- **Reduce the work of your working memory**
  Your working memory has a maximum capacity of information it can process. If that load is exceeded, learning does not take place. Visuals can help reduce the load and allow more brain power to be devoted to learning new material.
- **Stimulate your brain so that you enjoy learning more**
  Visual fireworks can liven up your experience of learning and therefore make information easier to recall.

Visual approaches to learning can include:

- Diagrams, photos, cartoons, posters, graphs and flowcharts of items to be learned

- Colour-coding and highlighting text material
- Visualizing concepts and sequences
- Flashcards

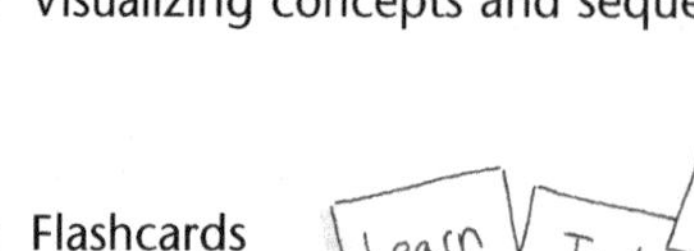

- Headings, sub-headings and bullet points to separate ideas
- Body/Mind mapping

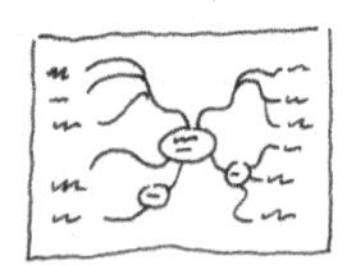

- Watching DVDs, films, computer animations or live demonstrations

Visuals are used extensively throughout this book and you will find more examples to inspire you in this chapter.

Activity: The digestive system

1. Study the digestive system using your favourite textbook or handouts. Check out anything you don't understand.
2. Look at the 3 different ways the digestive system is illustrated over pages 59 to 61.
   - in words, with ideas separated by bullet points and important points highlighted in colour;
   - with a simple diagram;
   - using a cartoon.
3. Use all 3 ways to revise the digestive system.
4. When you feel ready, draw a simple diagram to test yourself.

How effective was this method of revising?

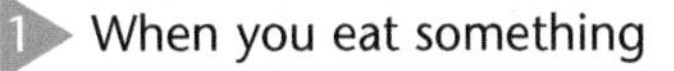

The Digestive System

1 When you eat something

- It is mashed up by your **teeth** and moved along by your **tongue** towards your **pharynx** (throat).
- The **salivary glands** produce saliva: this begins the digestion of starch and lubricates the descent of food.
- Food is moved by **peristalsis** (contraction of muscles) down your **oesophagus** to your **stomach**.

2 The **stomach**

- It churns and mixes the food with gastric juices: **pepsin, rennin** and **hydrochloric acid**. These chemicals break down the **fat** and **protein** and kill **bacteria**.
- The **pyloric valve** releases partially digested matter into the **small intestine**.

3 The **small intestine** consists of: **duodenum, jejunum** and **ileum**

- It receives **pancreatic juices** and **bile** to break down fats.
- It is lined with a mucous membrane covered with **villi**. Villi are 1mm finger-like projections containing **lacteals** for the absorption of fat into **lymph vessels** and **capillary loops** for the absorption of sugars and proteins into the blood.

4 **The large intestine**

- This consists of: the **ileo-caecal valve** (for onflow of intestinal contents); **caecum with appendix**; **ascending, transverse and descending colon**; **rectum** and **anus**.
- It receives undigested food – **cellulose**.
- It **absorbs water and salts into your blood**.
- **Used blood** goes to **kidneys** to be cleaned; clean blood goes back into the bloodstream.
- Remaining **waste products** are excreted via the rectum and anus.

5 The **liver** receives the products of digestion from the bloodstream. It produces, stores and releases vital substances: **vitamins, glycogen, bile, blood proteins**.

6 The **nutrients** produced from food – **vitamins, minerals, sugars, amino-acids** and **fatty acids** – are necessary for growth, repair, nourishment, energy and heat.

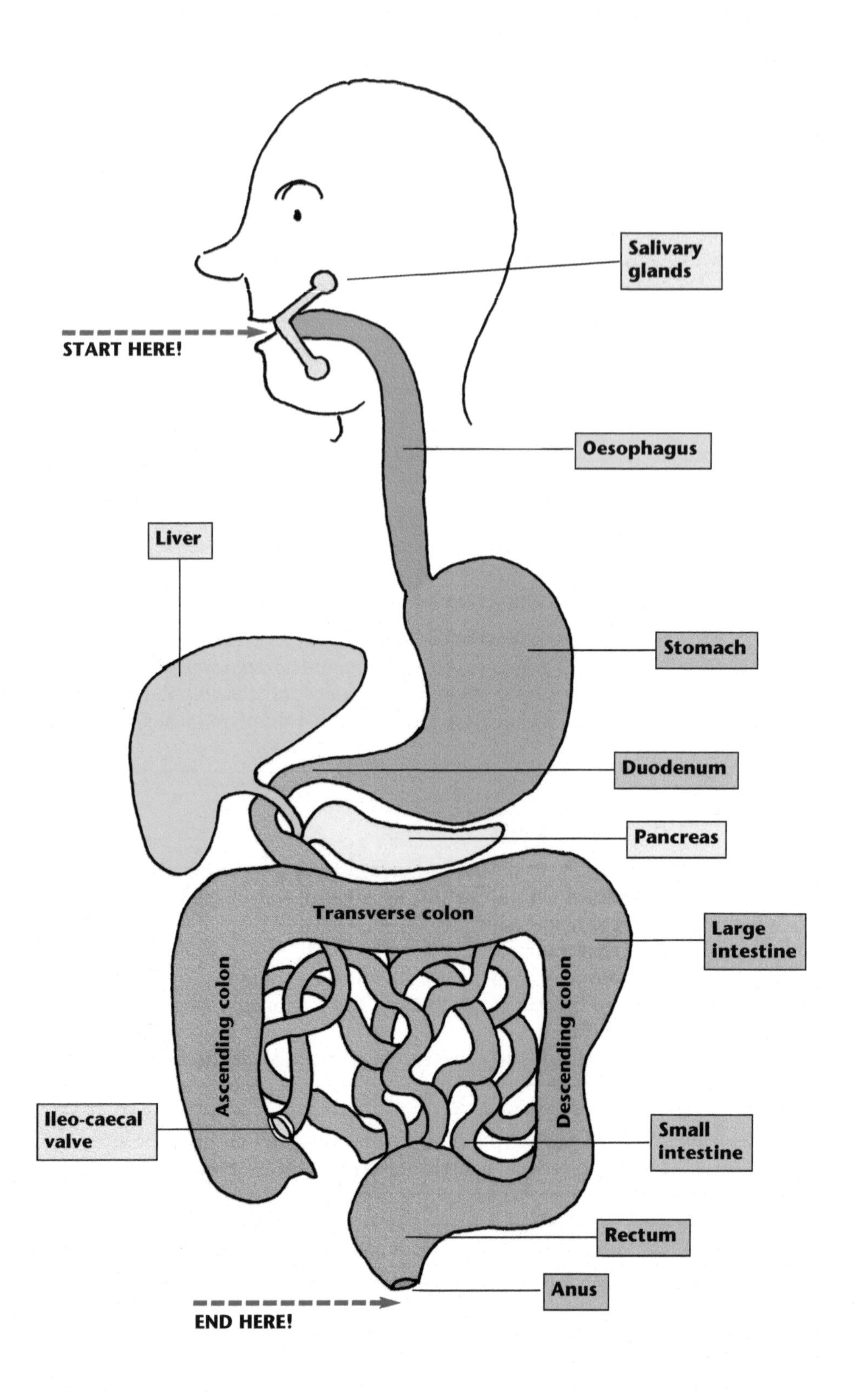
Salivary glands
START HERE!
Oesophagus
Liver
Stomach
Duodenum
Pancreas
Transverse colon
Large intestine
Ascending colon
Descending colon
Ileo-caecal valve
Small intestine
Rectum
Anus
END HERE!

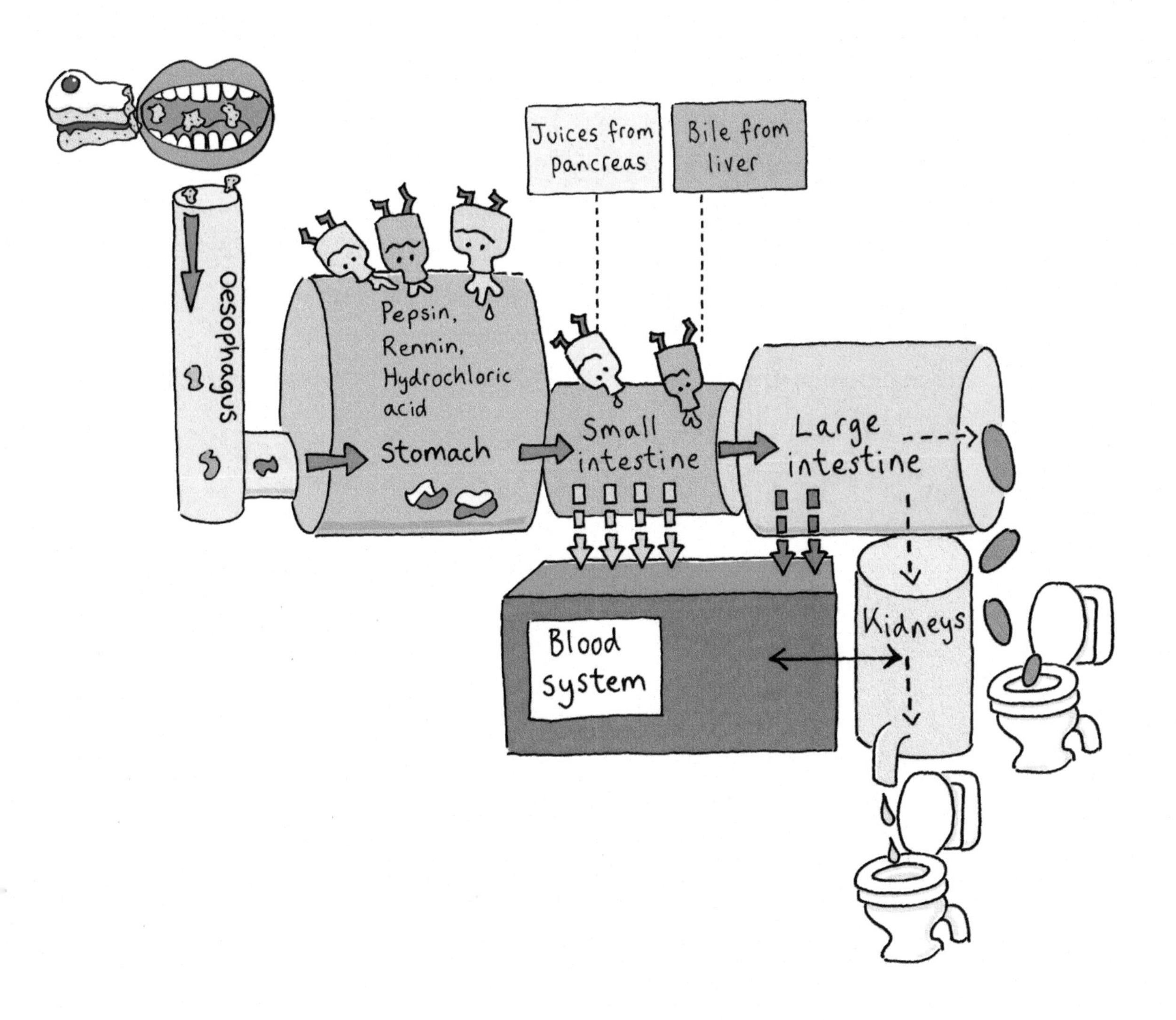

The Digestive System

I think that you have guessed that this diagram is not drawn to scale or anatomically correct! But it will help you remember the digestive process. Look at it as you read about the process on page 59.

Activity: Spot the errors

- Study the map carefully on page 63, and then cover it up so you can't see it.
- Correct the notes below: there are 10 errors in all.
- Then check back to the map to correct yourself.
- Now try to re-create the map or the bullet point notes from memory.
- Repeat this procedure until you no longer make any mistakes.

How successful is this revision method for you?

ADULT CASUALTY HAS COLLAPSED

Assess Situation

1 Danger to casualty?

- a. Ignore danger to self
- b. Move casualty – only if necessary

2 Check vital signs?

- a. Consciousness
  - Shake vigorously
  - Pinch casualty
  - Speak quietly
- b. Breathing
  - Tilt head forwards
  - Feel for breath on your hand
  - Listen for ambulance
  - Look for eye movement

3 Take Action

- a. No breathing?
  - Start chest compressions then phone 999
- b. Breathing?
  - Phone 999 immediately
  - Basic Life support

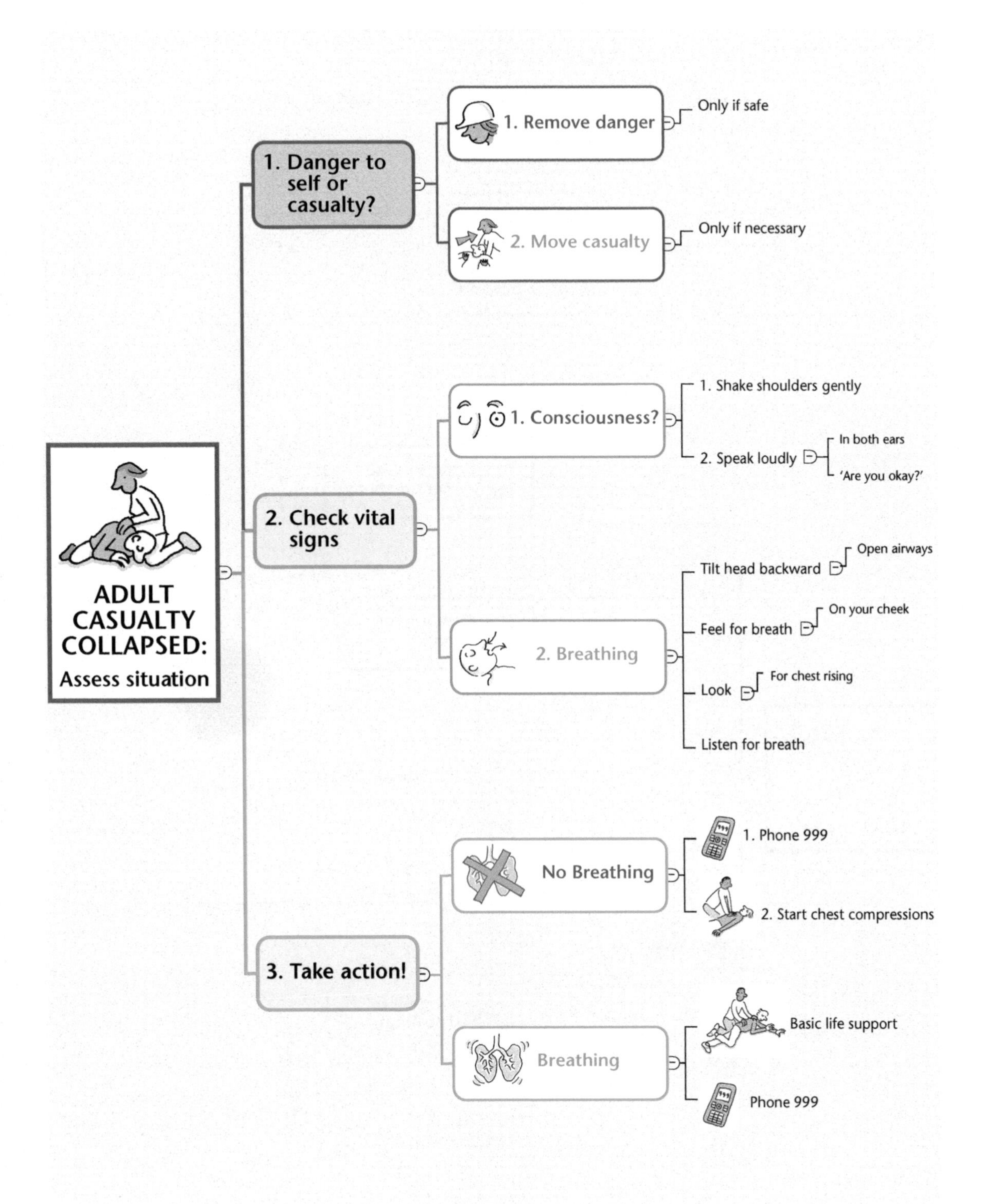

ADULT CASUALTY COLLAPSED:
Assess situation
1. Danger to self or casualty?
1. Remove danger
Only if safe
2. Move casualty
Only if necessary
2. Check vital signs
1. Consciousness?
1. Shake shoulders gently
2. Speak loudly
In both ears
'Are you okay?'
2. Breathing
Tilt head backward
Open airways
Feel for breath
On your cheek
Look
For chest rising
Listen for breath
3. Take action!
No Breathing
1. Phone 999
2. Start chest compressions
Breathing
Basic life support
Phone 999

Activity: Spot the Errors

1. Study carefully the 'Unconscious' map on page 65.
2. Then cover it up, so you can't see it.
3. Correct the map below: there are 10 errors in all.
4. Next, check back to the map on page 65 to correct yourself.
5. Finally, try to sketch a quick map by yourself from memory.
6. Repeat these procedures until you no longer make any errors.

How effective was this method for revising?

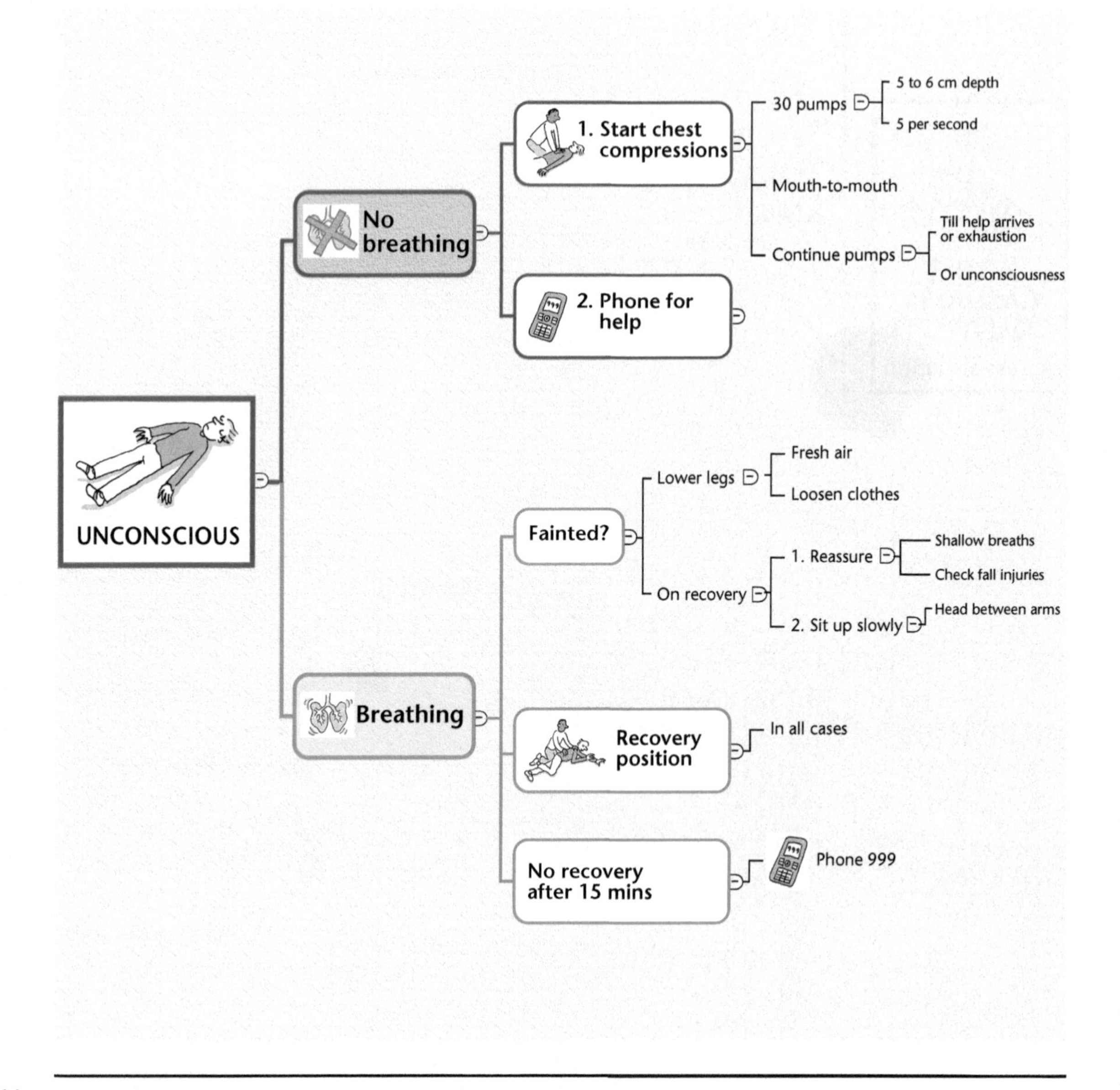

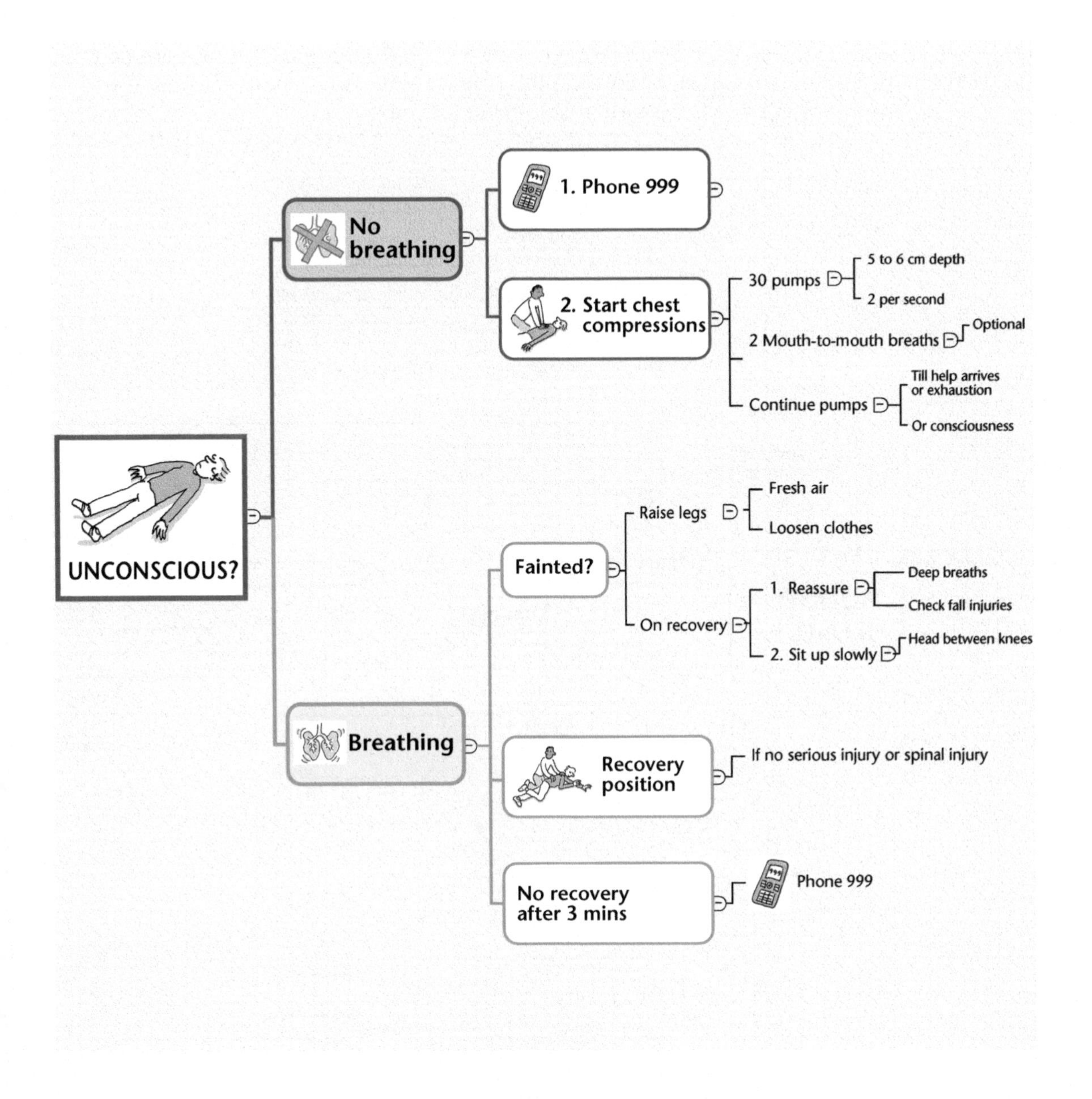
UNCONSCIOUS?
No breathing
1. Phone 999
2. Start chest compressions
30 pumps
5 to 6 cm depth
2 per second
2 Mouth-to-mouth breaths
Optional
Continue pumps
Till help arrives or exhaustion
Or consciousness
Breathing
Fainted?
Raise legs
Fresh air
Loosen clothes
On recovery
1. Reassure
Deep breaths
Check fall injuries
2. Sit up slowly
Head between knees
Recovery position
If no serious injury or spinal injury
No recovery after 3 mins
Phone 999

Activity: Sudden illness Fill in the gaps

Look at the Map of Sudden Illness on page 67. Cover it up and fill in the gaps on this map. Then test yourself by covering this map up and quickly re-drawing your own map. How effective is this system of learning for you?

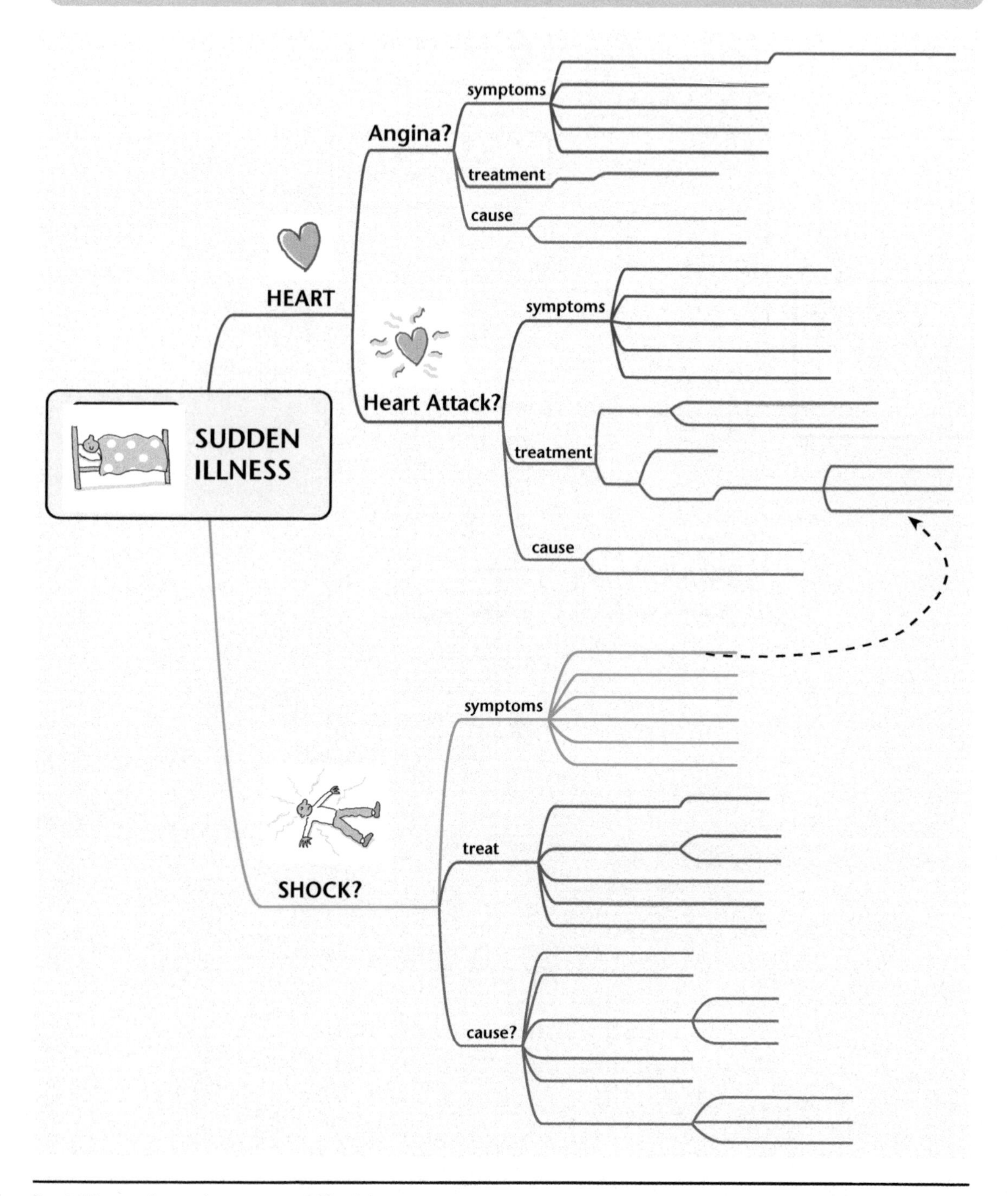

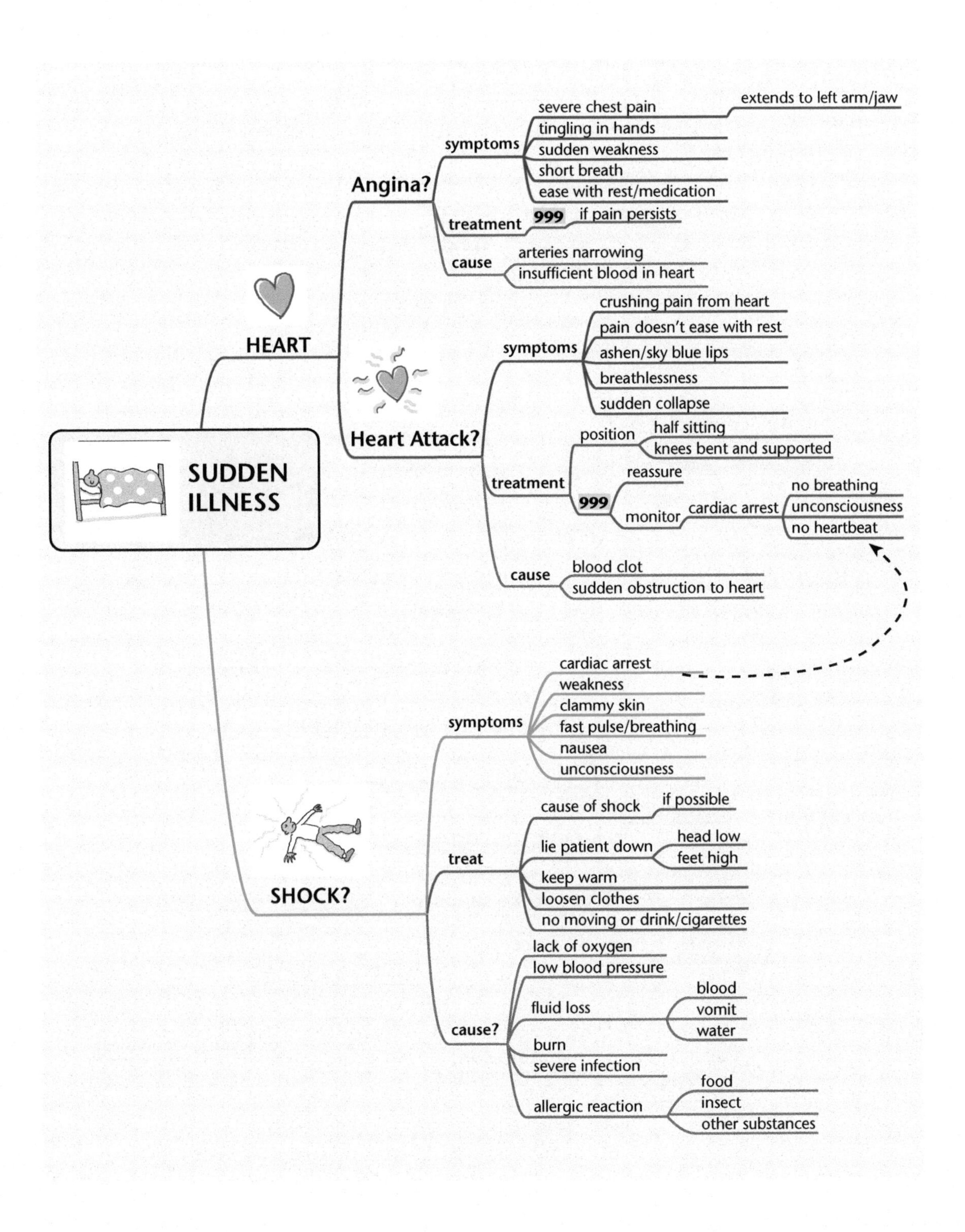
SUDDEN ILLNESS
HEART
Angina?
symptoms
severe chest pain
extends to left arm/jaw
tingling in hands
sudden weakness
short breath
ease with rest/medication
treatment
999
if pain persists
cause
arteries narrowing
insufficient blood in heart
Heart Attack?
symptoms
crushing pain from heart
pain doesn't ease with rest
ashen/sky blue lips
breathlessness
sudden collapse
treatment
position
half sitting
knees bent and supported
999
reassure
monitor
cardiac arrest
no breathing
unconsciousness
no heartbeat
cause
blood clot
sudden obstruction to heart
SHOCK?
symptoms
cardiac arrest
weakness
clammy skin
fast pulse/breathing
nausea
unconsciousness
treat
cause of shock
if possible
lie patient down
head low
feet high
keep warm
loosen clothes
no moving or drink/cigarettes
cause?
lack of oxygen
low blood pressure
fluid loss
blood
vomit
water
burn
severe infection
allergic reaction
food
insect
other substances

Activity: Accident: Improve this map

Look at the map of First Aid for Injuries on page 69. Now colour-code, add symbols, extra information and pictures to make this map more personal and visually memorable.

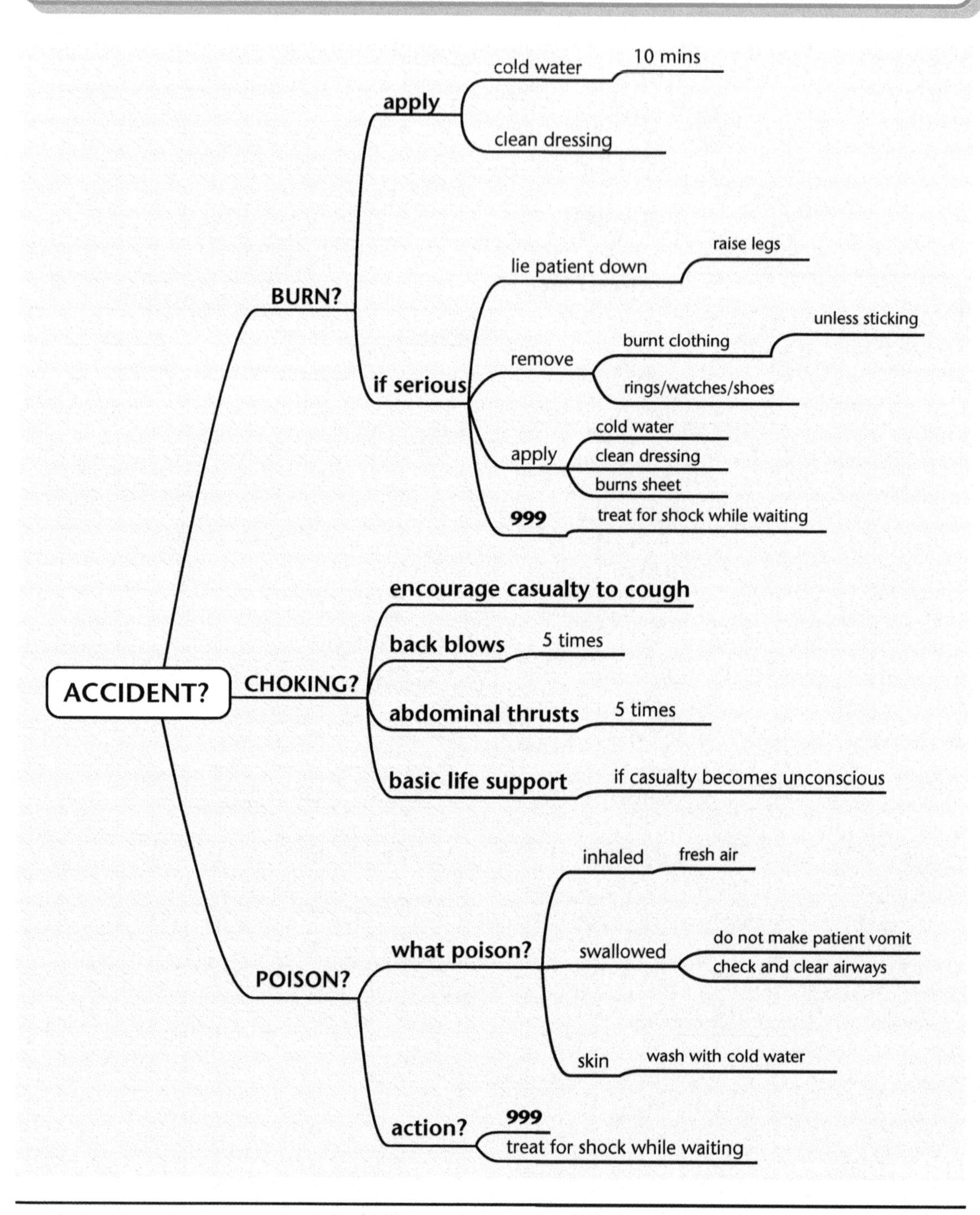

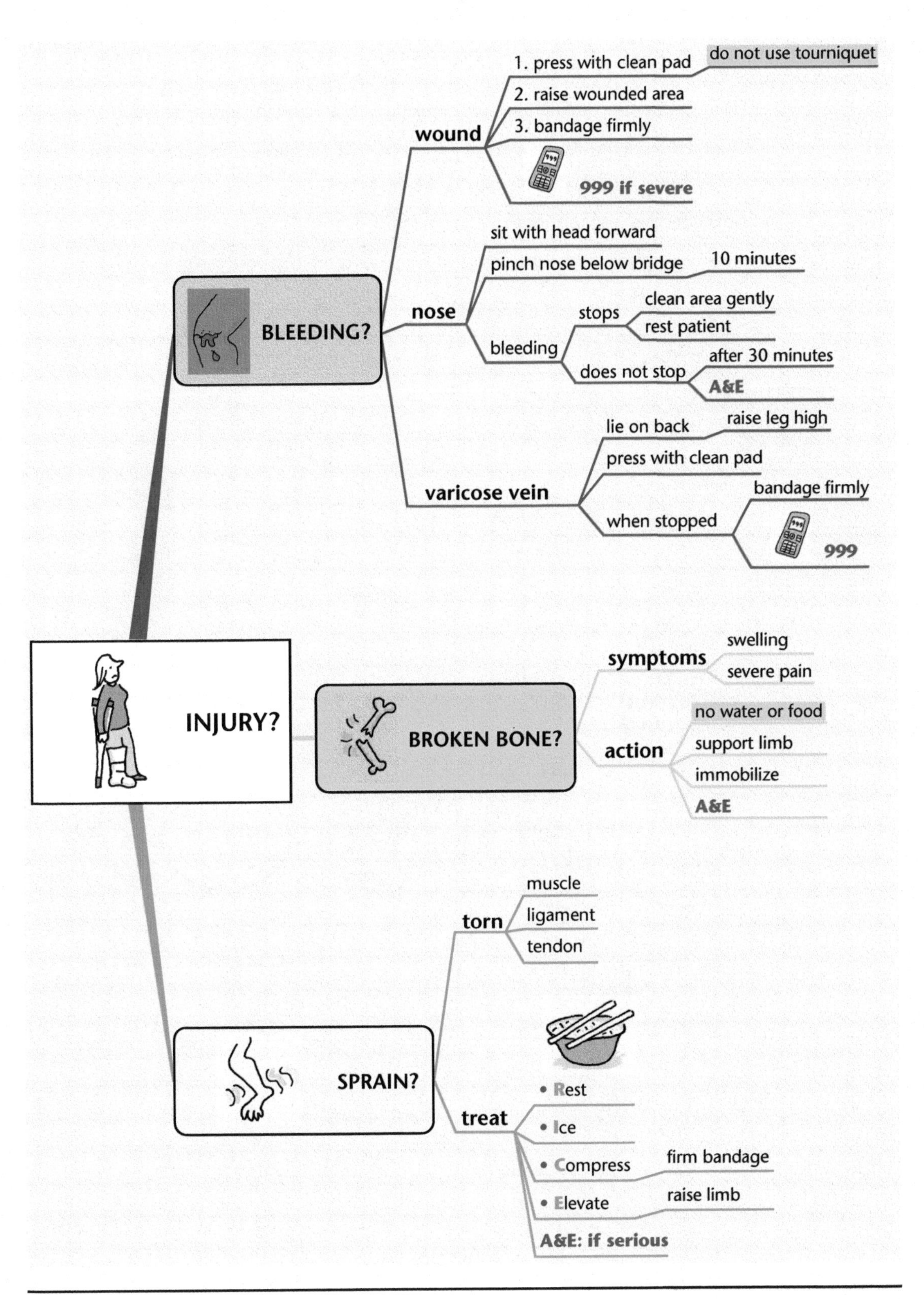
INJURY?
BLEEDING?
wound
1. press with clean pad
do not use tourniquet
2. raise wounded area
3. bandage firmly
999 if severe
nose
sit with head forward
pinch nose below bridge
10 minutes
bleeding
stops
clean area gently
rest patient
does not stop
after 30 minutes
A&E
varicose vein
lie on back
raise leg high
press with clean pad
when stopped
bandage firmly
999
BROKEN BONE?
symptoms
swelling
severe pain
action
no water or food
support limb
immobilize
A&E
SPRAIN?
torn
muscle
ligament
tendon
treat
• Rest
• Ice
• Compress
firm bandage
• Elevate
raise limb
A&E: if serious

Activity: Heat exhaustion

- Fill in the map on the left without looking at the answers on the right.
- Then check your answers are correct. Repeat till perfect!
- Finally try to sketch a quick map by yourself.
- How effective was this method for revising?

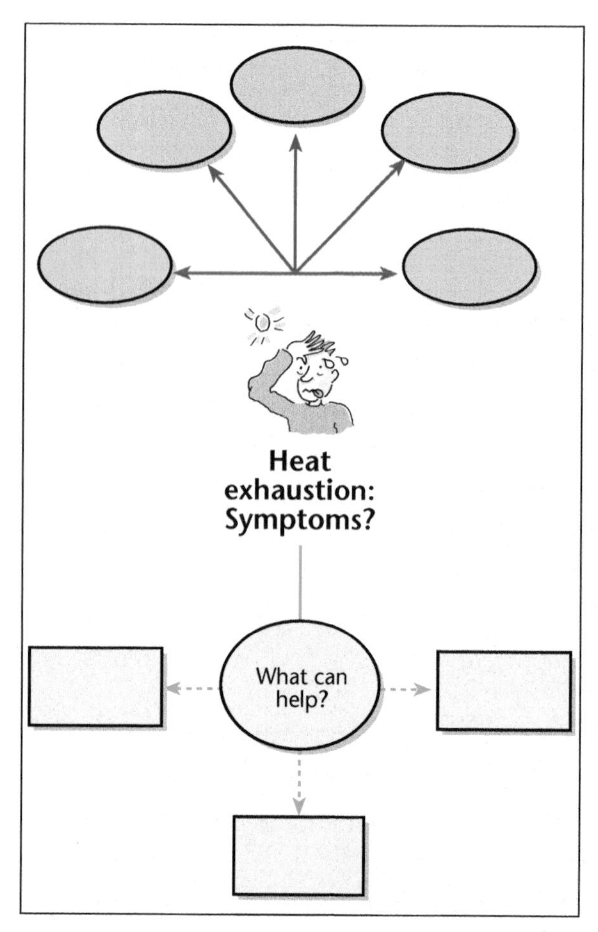

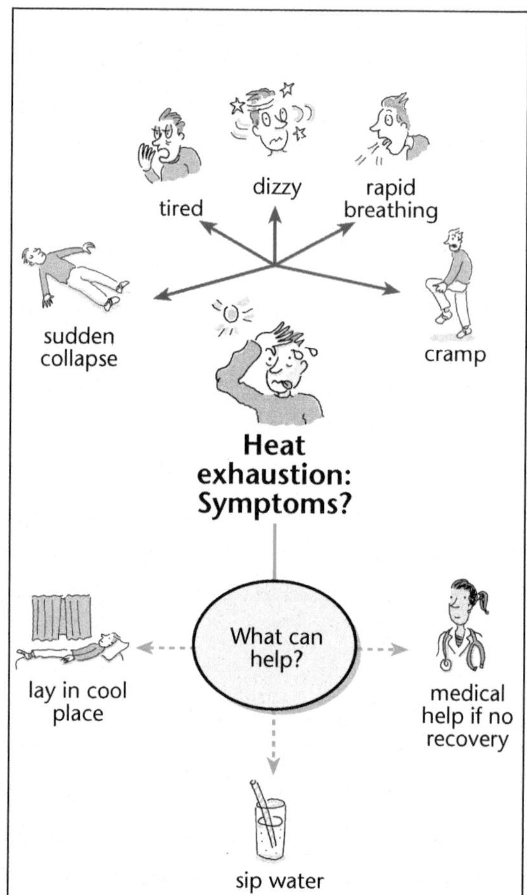

Activity: Panic attack

1. Study the map below, then cover it up.
2. Sketch a quick map by yourself.
3. Compare your map with the original one below and note any missing parts.
4. Repeat procedure until your map has all the correct information.

How effective was this method for revising?

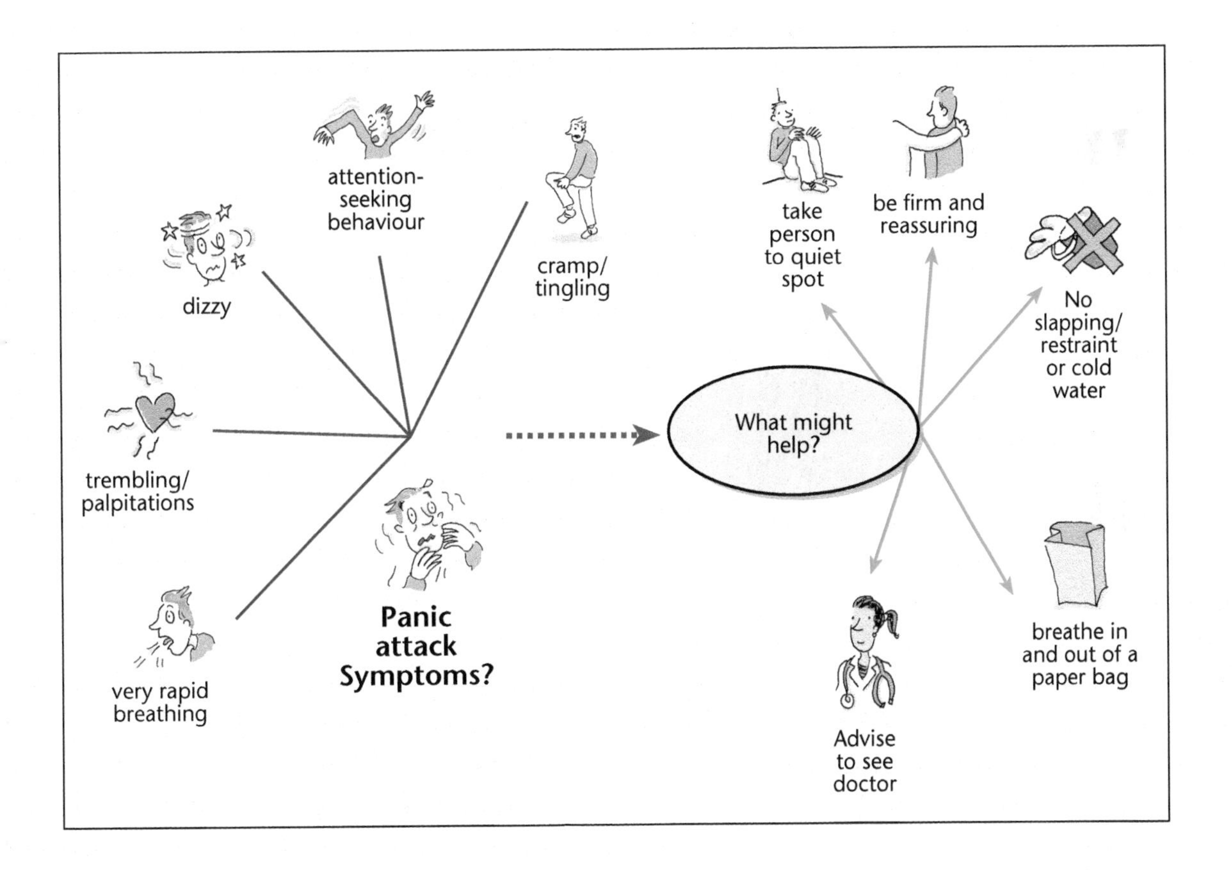

Colour-coding

This strategy is used extensively throughout this book, and is a very important resource for you. See examples on pages 30, 44, 117 and 144. Colour is processed in a particular area of the brain and is therefore an extra way to help fix information in your memory. Basically, you can use colour to **colour-code** similar ideas or highlight ideas that are different or important.

Activity: Colour-coding arm muscles

In the diagram below:

- Colour in **blue** the 2 muscles that extends the **wrist**:
  the **Extensor Carpi Ulnaris** and
  the **Extensor Carpi Radialis**.
- Then colour in **red** the **Extensor Digitorum**, which extends the fingers.

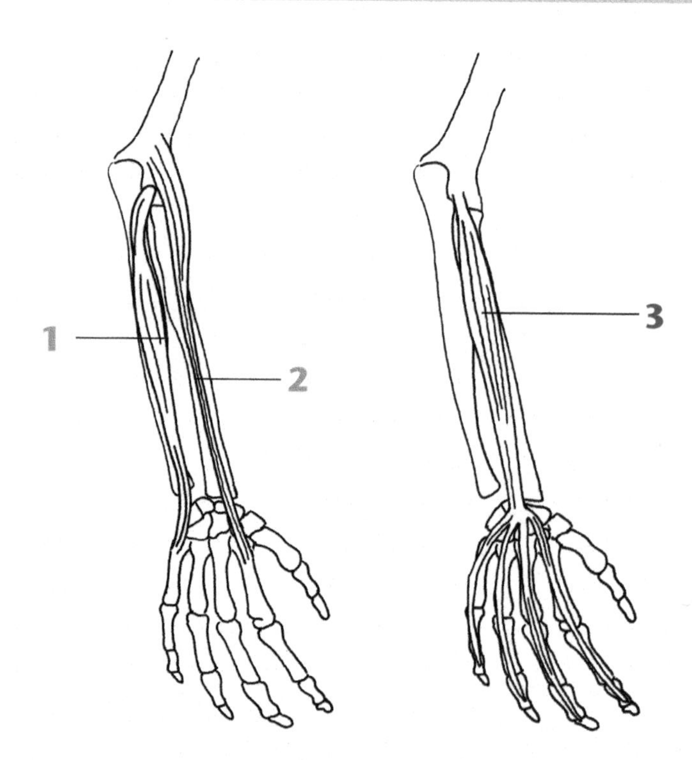

ARM MUSCLES – Posterior view

1 **Extensor Carpi Ulnaris:** **extends wrist**
2 **Extensor Carpi Radialis:** **extends wrist**
3 **Extensor Digitorum:** **extends fingers**

Flash cards

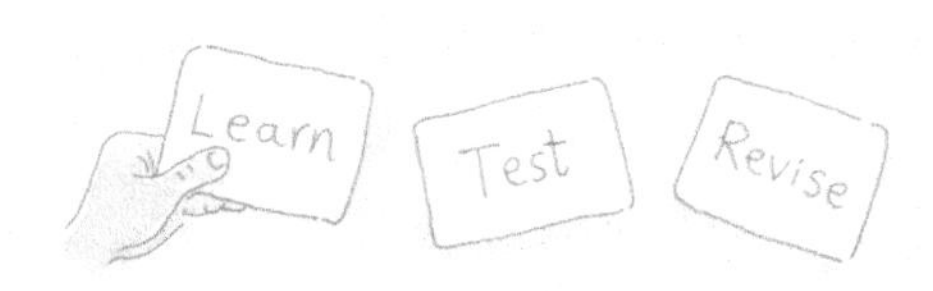

What is a flash card?

A flash card is literally a card (usually a 5 inch by 8 inch index card size) with a small piece of information on one side – a picture or a few key words – and a test question on the other side. Basically, it is used to flash your brain with an image. In order for a flash card to be useful, it's not enough to just glance at it. You will need a Learn–Test–Revise strategy that involves engaging with the material, self-testing and then revising until recall becomes automatic. This is fully illustrated on page 74.

Why do we forget information?

Having trouble remembering something is nothing to do with intelligence. The psychologist Hermann Ebbinghaus came up with some clever experiments to work out why we remember or forget information. He found that without **repetition** or **encoding** the memory traces decayed rapidly in the brain. In fact, we usually lose about 75% of what we learn after only 48 hours without special encoding. This would be depressing if there was nothing you could do about it.

What does encoding mean?

Encoding means transforming something you see, hear, think, smell, feel or do into a memory. There are two different forms of encoding:

1. *Shallow encoding* involves repeating information to yourself over and over again (such as a phone number). Shallow encoding is useful when we want to hold a small amount of information in mind for a few seconds. This system uses part of the brain designed only for short-term memory purposes and so the information is only fixed temporarily. The brain is constantly flooded with information from our senses and so has to get rid of most of the information that it does not need.

2. *Elaborative encoding* involves deliberately connecting the new information with something already in your memory in order to make it meaningful. This is the best way to fix information into long-term memory – where things are stored for use later.

So how can you avoid forgetting information which you have just studied?

The simple trick is to repeat the information neither too early nor too late. This is explained below.

Following on from Ebbinghaus's research, the psychologist Sebastian Leitner devised the 'Card-box Method' for learning information with flash cards. The method is based on simple rules and makes use of a card-box divided into sections. As each flash card has been successfully learned, it is promoted to the next section. This way you don't waste your time: things that are difficult to learn are presented more often for memorization, and those that have been driven into long-term memory are repeated far less frequently.

Below you will find a modern version of the Card-box Method, which is based on sound research and which has been found by other students to be extremely effective.

You will need an index box divided into 8 sections

- NEW CARDS
- TODAY
- SAME DAY
- 24 HOURS LATER
- 3 DAYS LATER
- 1 WEEK LATER
- 1 MONTH LATER
- LEARNT CARDS

ROUTINE TO FOLLOW:

1. A maximum of 10 new cards are picked each week for learning, from a pool of cards (in the NEW CARDS section). You can add to these new cards whenever you want. Put the 10 cards into the section marked TODAY.
2. Take one card at a time from the TODAY box and study it carefully.
3. Turn over the card and test yourself.
4. If you have made an error, have another go. If you are correct, put the card into the SAME DAY section.
5. Then repeat with the next card until you have revised all 10 cards and all 10 are in the SAME DAY section.
6. If you are still making errors on a card at the end of the week, that card goes back into the TODAY box for another week!

To drive the information into your long-term memory, you will need to repeat steps 1 to 6 above:

- **Later on the same day**
- **24 hours later**
- **3 days later**
- **1 week later**
- **1 month later**
- **then file in the LEARNT CARDS section**

Leitner found that a card that was successfully repeated 5 times in a row has reached long-term memory. But everyone's learning capacity is different, so don't get discouraged – just stick to the method. It will work for you.

Place your cards where you can see them frequently: for example, on the fridge, above your desk, or inside your bag. Alternatively, there is some brilliant computer software to help you create electronic flash cards. See the 'Useful resources' section page 171.

Activity: Making flash cards for revision

- Copy the following images of the brain onto index cards.
- Label, and colour in, the different parts of the brain.
- Copy out the definitions on the reverse of the cards.
- Use the cards for testing yourself or other students.

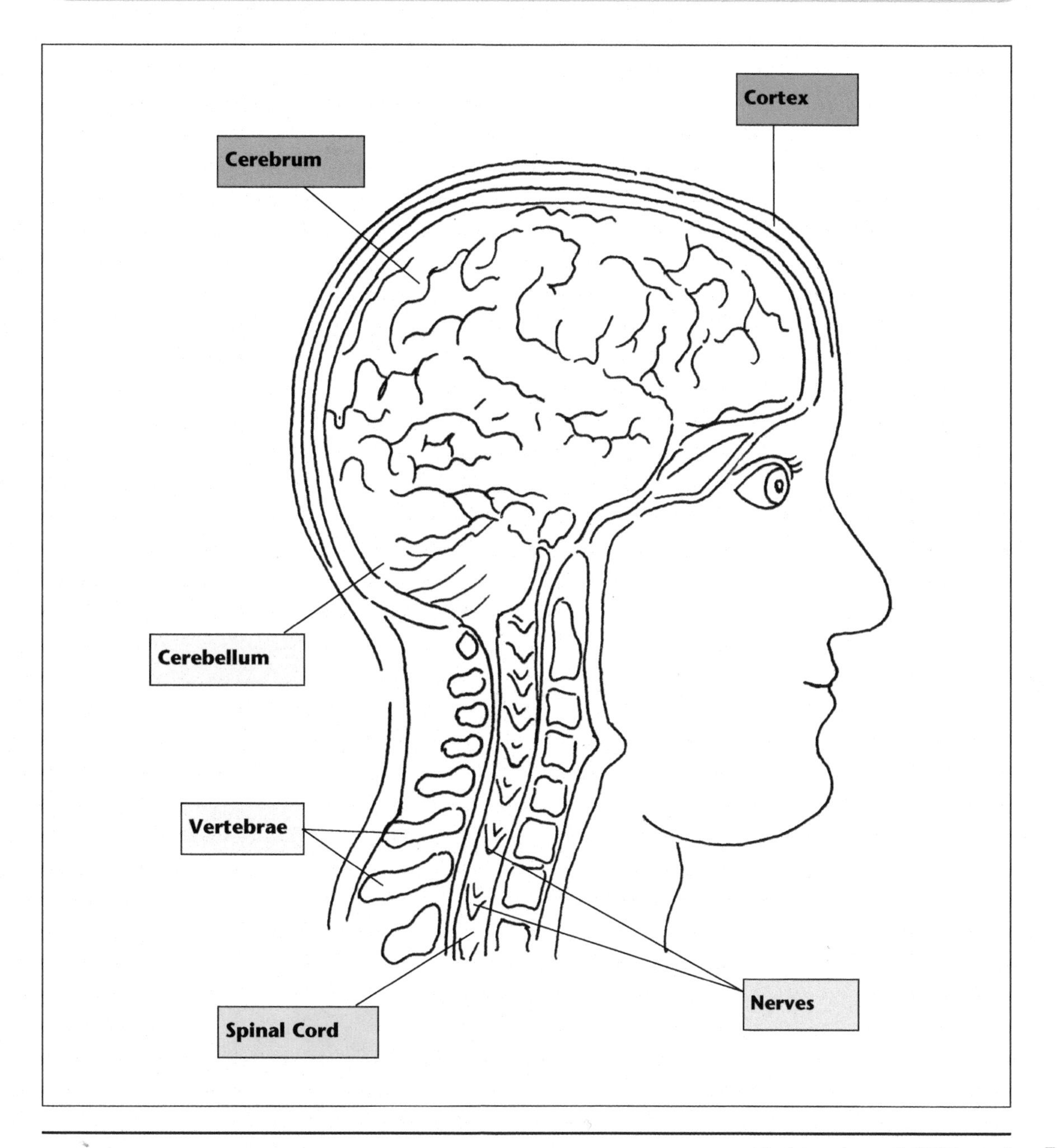

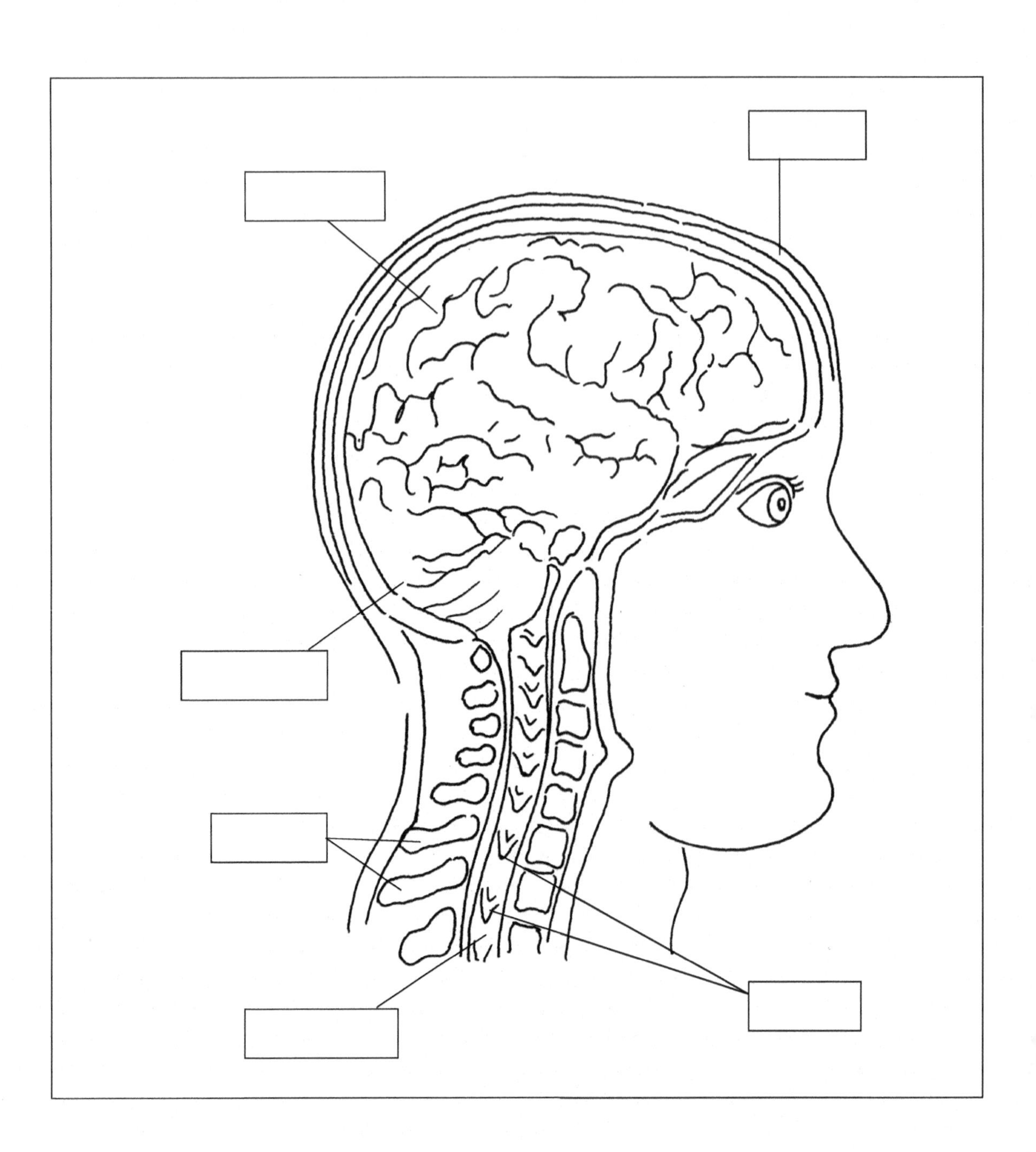

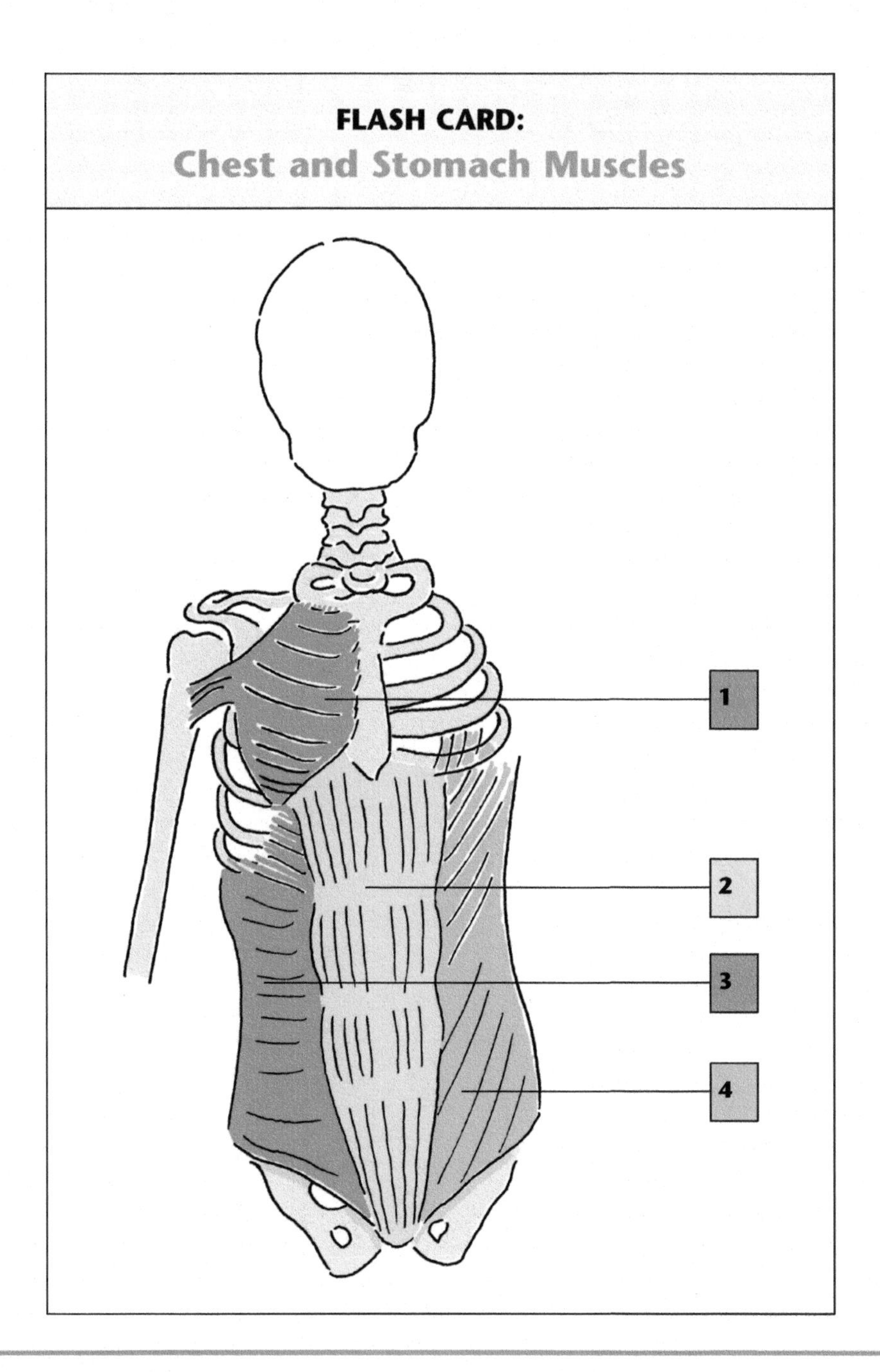

Copy onto back of card:

1 **Pectoralis**: flexes, adducts and medially rotates **humerus**
2 **Rectus Abdominis**: flexes trunk and compresses **abdomen**
3 **Abdominis Transversalis**: compresses **abdomen**
4 **External Oblique**: laterally flexes and rotates **trunk** and compresses **abdomen**

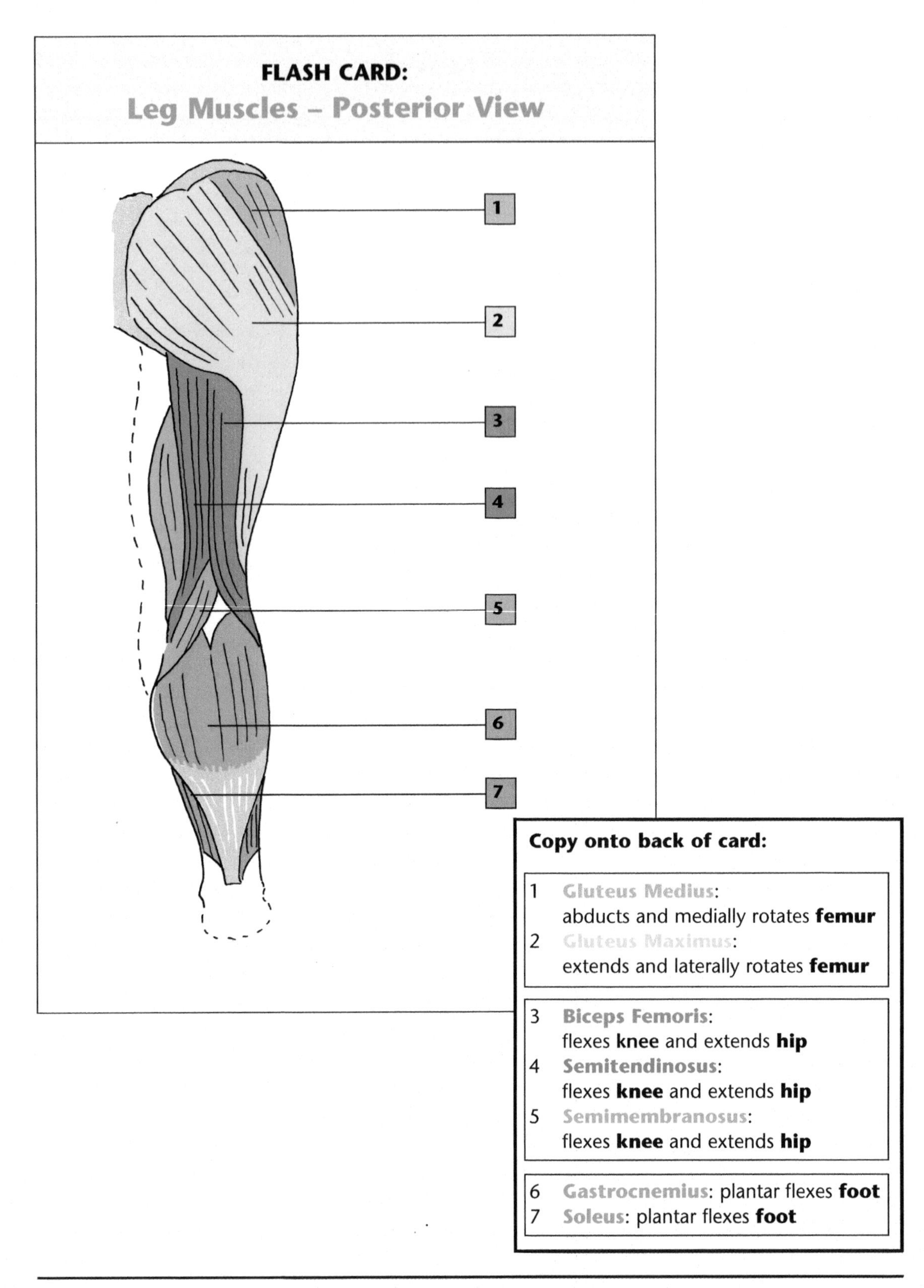
FLASH CARD:
Leg Muscles – Posterior View
1
2
3
4
5
6
7
Copy onto back of card:
1 Gluteus Medius: abducts and medially rotates femur
2 Gluteus Maximus: extends and laterally rotates femur
3 Biceps Femoris: flexes knee and extends hip
4 Semitendinosus: flexes knee and extends hip
5 Semimembranosus: flexes knee and extends hip
6 Gastrocnemius: plantar flexes foot
7 Soleus: plantar flexes foot

Activity: Bookmarks for revision

- Make a simple diagram and key notes on a piece of card the shape of a bookmark.
- You can use it to mark a particular topic and keep it ready as a quick flash card for revision.
- Look at the examples below.

Muscles of Forearm and Hand – anterior view

1 Brachialis: **flexes elbow**
2 Brachioradialis: **flexes elbow**

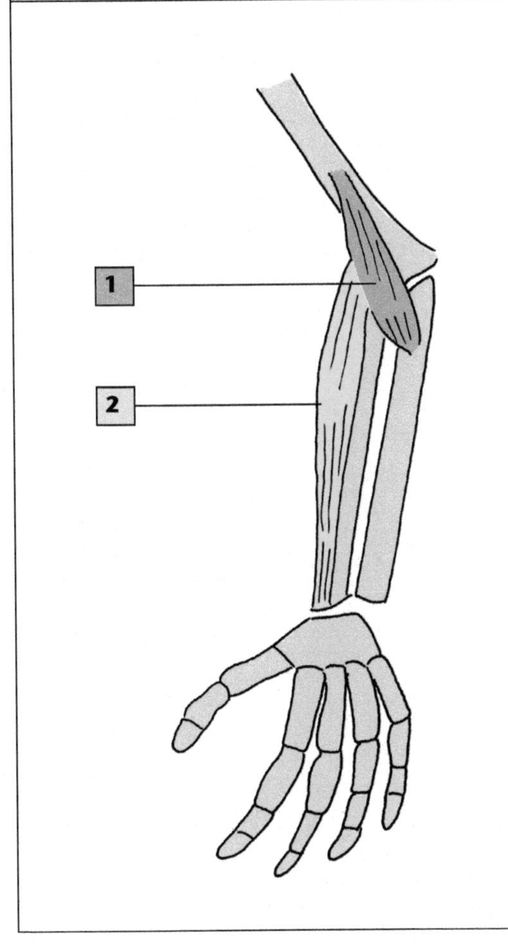

Muscles of Forearm and Hand – anterior view

3 Pronator Teres: **flexes elbow**

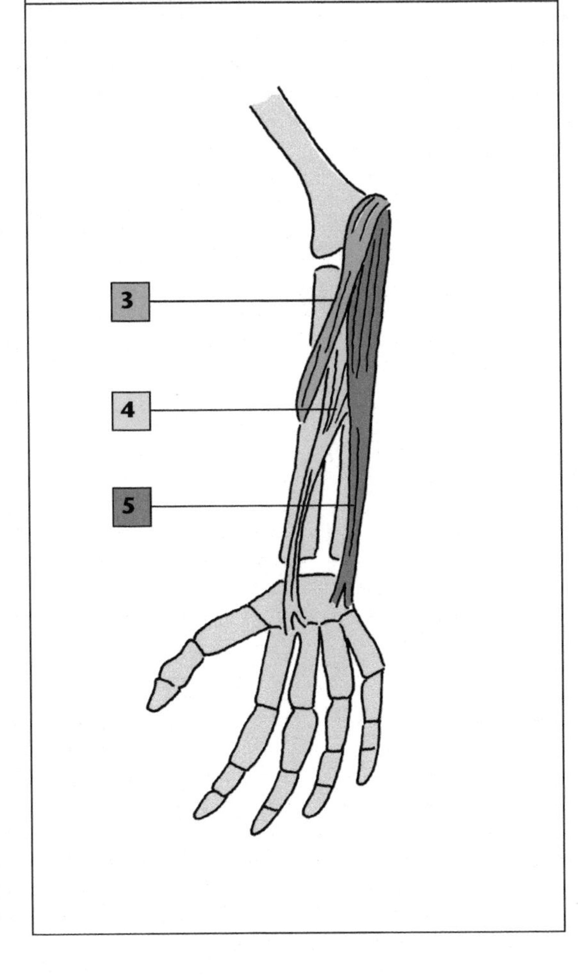

Muscles of Forearm and Hand – anterior view

4 Flexor Carpi Radialis: **flexes** wrist
5 Flexor Carpi Ulnaris: **flexes** wrist

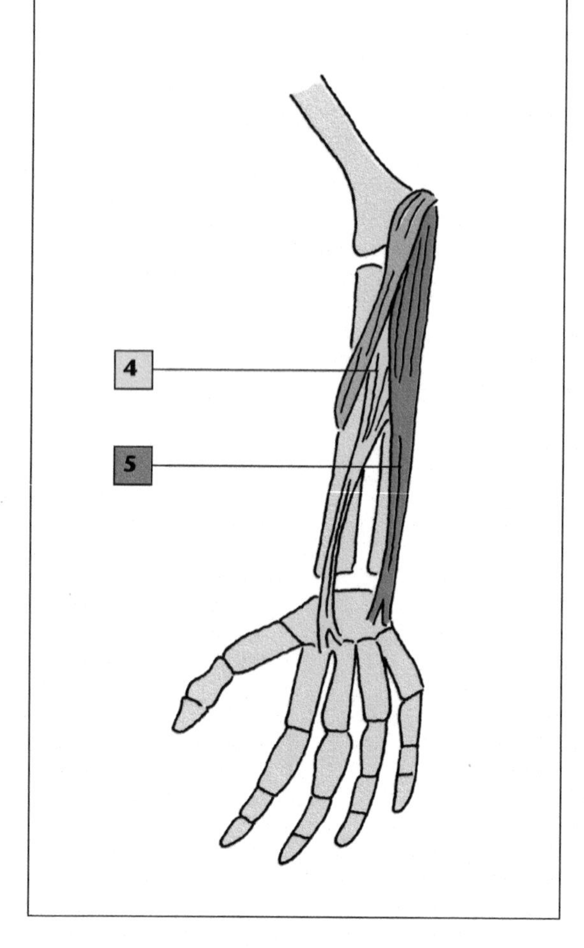

Muscles of Forearm and Hand – anterior view

6 Flexor Digitorum: **flexes** fingers

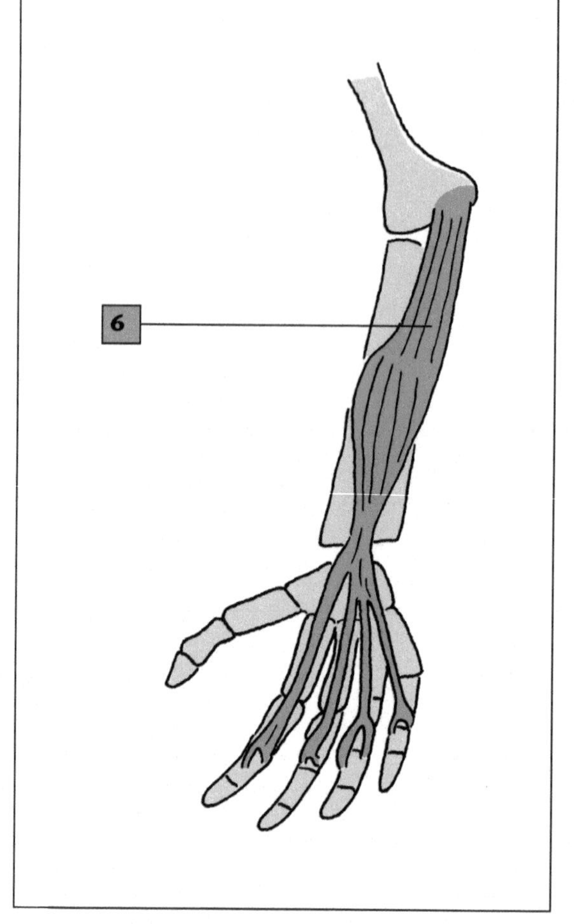

Chapter 5

Use all your senses

OVERVIEW

In this chapter, you will:

- ✓ **explore the potential of Sensory Learning;**
- ✓ **understand how to use your senses (such as movement, hearing, visualizing . . .) to make learning easier;**
- ✓ **get plenty of opportunities to practise this effective learning technique.**

Sensory learning

A smell, a photo, a taste, a few notes from a song, the movement of a dance, the feel of a piece of fabric can instantly set off a flood of memories from the past. This is the secret of a very powerful memory technique. You are far more likely to remember information if:

- you can deliberately fix whatever you want to remember to a sound, an image or a sensation;
- you process it in several different ways – sight, sound and movement.

These strategies build up very strong memory traces in several sites of the brain – the more the merrier. For example, if you need to fix something in your mind, you can use a combination of the following:

- watch a CD or computer animation;
- draw a cartoon;
- visualize the process in your own body;
- colour in the different parts of a diagram;

- make a Bodymap;
- learn it to music;
- explain it to somebody else;
- read it aloud to yourself or record it;
- reorganize your notes in different ways;
- write out lecture notes in your own words;
- read about it;
- label a flow chart/diagram.

Activity: Learning the names of body movements and positions

1. Make yourself a poster from the diagrams on page 83. You may want to do this by increasing the size to A3 on a photocopier.
2. Look at the pictures one by one, and practise each move, repeating its name **at the same time**. Remember: **synchronizing** sight, movement and sound is important.
3. If you record your own voice and play it back, you will set down a very strong memory trace. So:
   - record the names of the different body positions onto a recording device/MP3 player. Remember to leave a long enough interval between each word, so that you are able to practise the movement several times. You might also like to record some music in the background.
   - play your recording as often as you need to, and give your mind and body a good workout. The more exaggerated the movement, the more likely you are to remember it . . . Does this sound weird to you? Never mind! This technique really works.
4. Stick up your poster somewhere where you will often see it.
5. Get together with other students to practise the moves and test each other on the names.

Body movements and positions

Shoulder

abduction

adduction

flexion

extension

medial rotation

lateral rotation

Elbow

flexion

extension

Forearm

supination

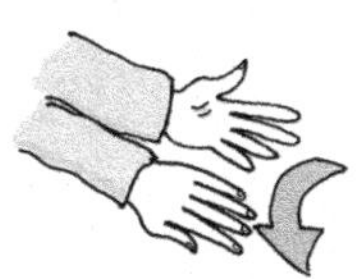
pronation

Wrist

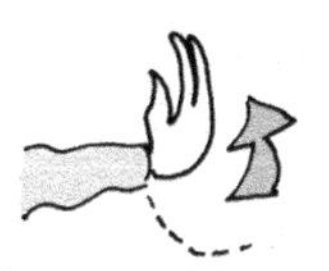
flexion

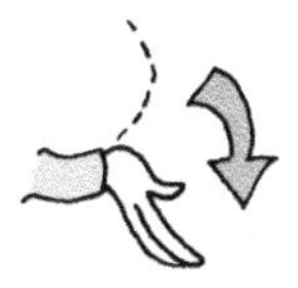
extension

Trunk

flexion

extension

lateral flexion

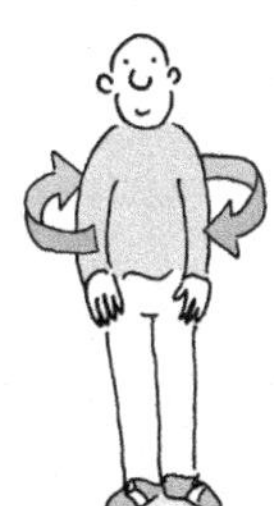
rotation

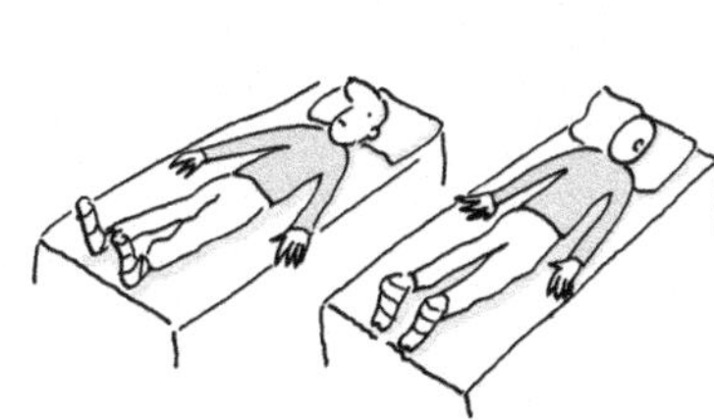
supine

prone

Hip

abduction

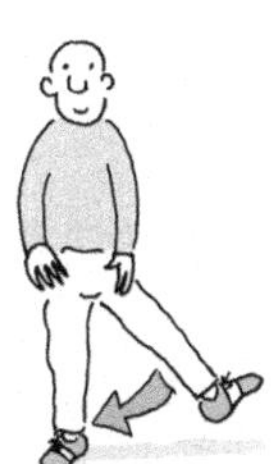
adduction

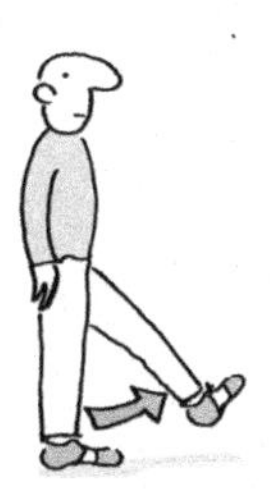
flexion

medial rotation

lateral rotation

circumduction

Knee

extension

flexion

Ankle

plantar flexion

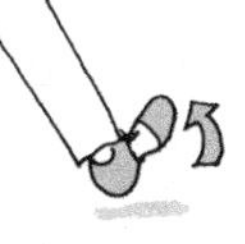
dorsi flexion

Activity: The respiratory system

Look carefully at the flowchart of the respiratory system. Now cover it up and practice labelling the blank chart on the next page in pencil.

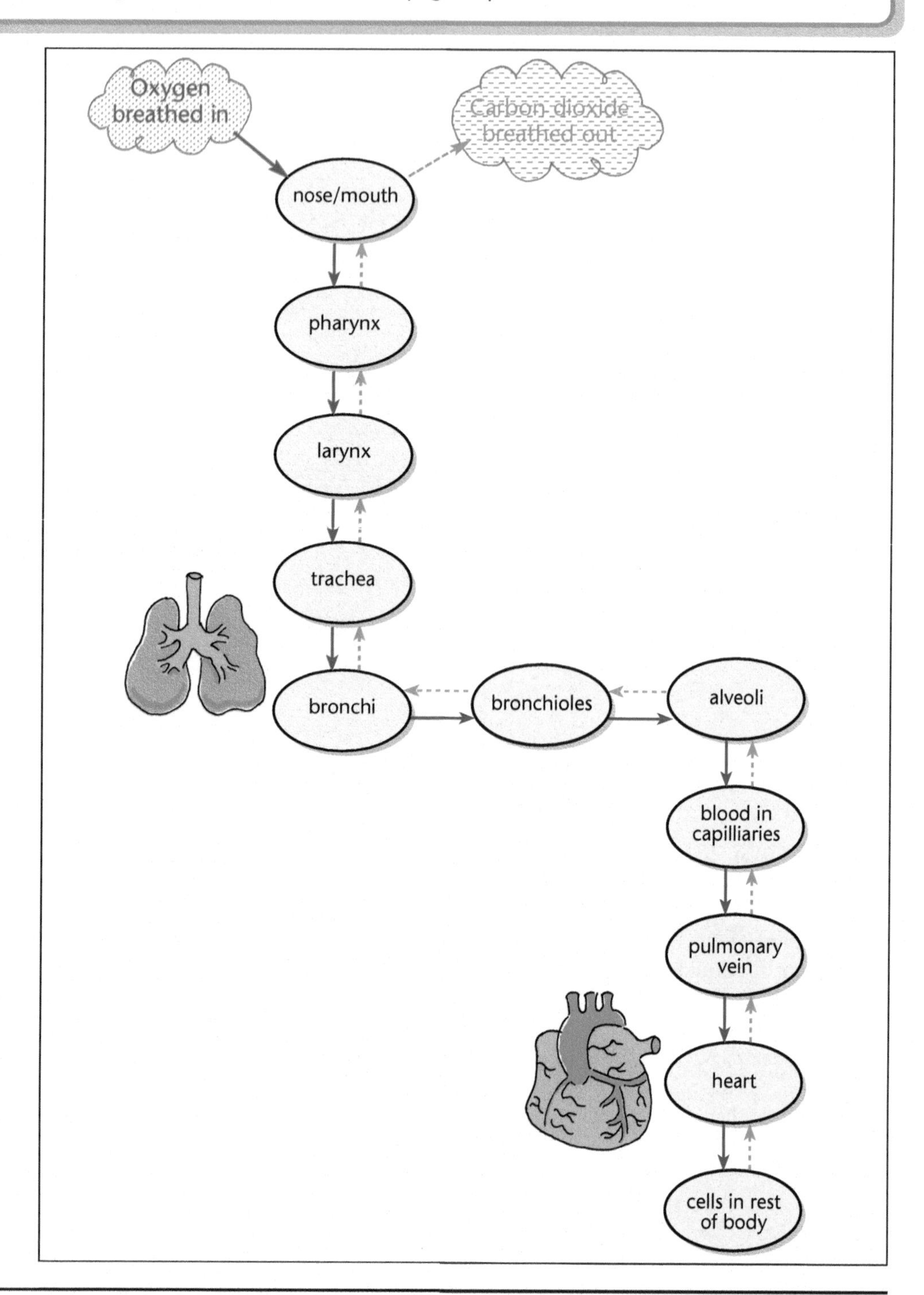

Diagram created in Inspiration® by Inspiration Software®, Inc.

Practise labelling this flowchart of the respiratory system in pencil. Practise till perfect using the 'Re-draw and Compare Method' (see page 36). This will fix the information in your memory.

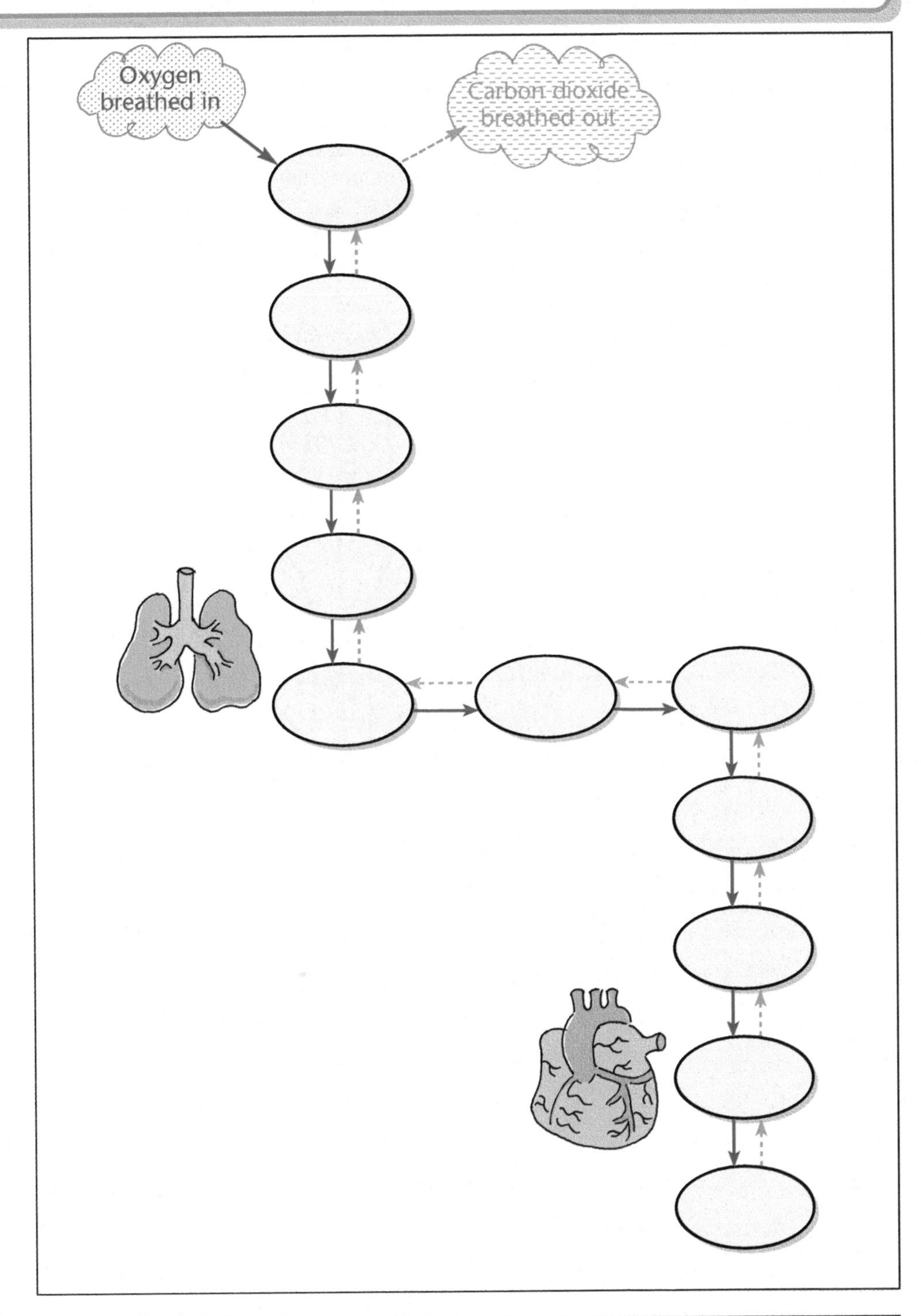

Activity: Posterior muscles

Training your auditory, visual and motor memories to work together is a very powerful learning technique. Can you write the names of the **Posterior Muscles** in pencil on this simplified diagram? The answers are on page 87.

Alternatively, it is extremely useful to photocopy this blank template onto transparent plastic and write on it with water-soluble felt tips that rub or wash off. You can use this template over and over again until your exams are over. See pages 88–9 for examples of templates for the posterior and anterior muscles.

You can make these templates for all the subjects that you need to revise.

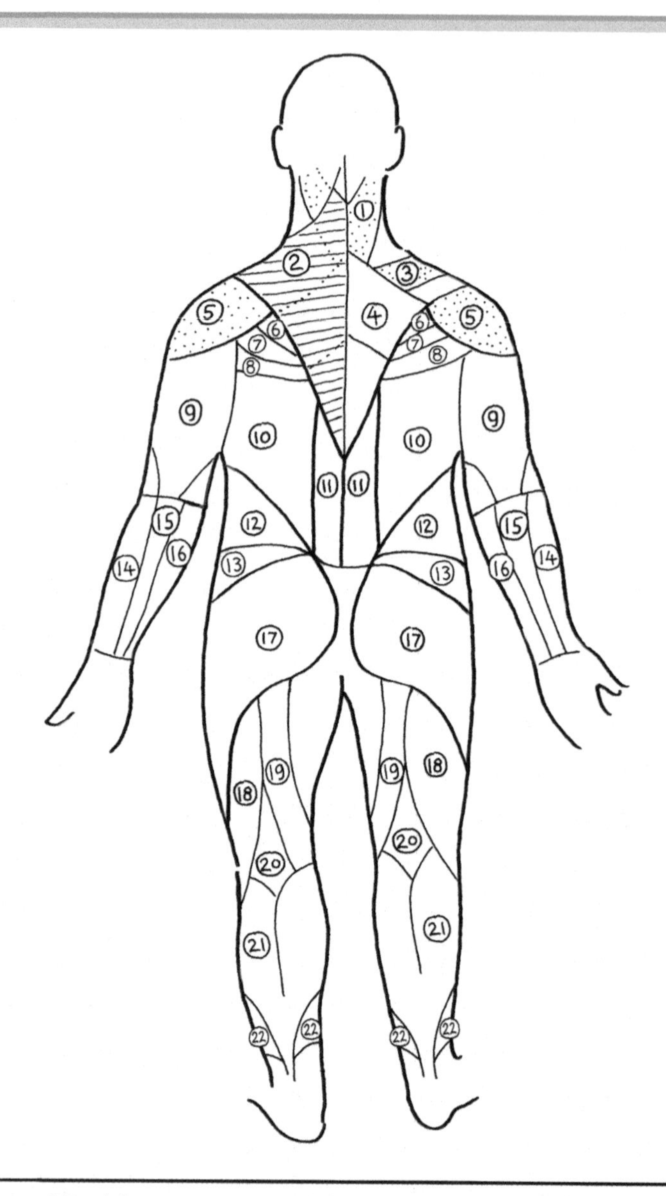

22 POSTERIOR MUSCLES

Note that most of the muscles on the left e-x-t-e-n-d!

1 Splenius capitis
- Extends head

2 Trapezius
- Extends head
- Lifts/draws back scapula

5 Deltoid
- Extends
- Flexes
- Abducts arm

9 Triceps
- Extends elbow

10 Latissimus dorsi
- Extends
- Adducts
- Medially rotates humerus

11 Erector spinae
- Extends spine

14 Extensor carpi radialis
15 Extensor carpi digitorum
16 Extensor carpi ulnaris
- Extend wrist

17 Gluteus maximus
- Extends
- Laterally rotates femur

3 Supraspinatus
- Abducts arm

4 Rhomboid
- Draws back scapula

6 Infraspinatus
7 Teres minor
- Both laterally rotate humerus

8 Teres major
- Medially rotates humerus

12 External obliques
- Flex and
- Rotate trunk

13 Gluteus medius
- Abducts and
- Medially rotates femur

18 Biceps femoris
19 Semitendinosus
20 Semimembranosus
- Flex knee
- Extend hip

21 Gastrocnemius
22 Soleus
- Flex foot

Some students find it very useful to cut out illustrations of muscles and practise sticking them on to somebody's body. They illustrations have to be life-sized and run from the origin to the insertion.

Template: posterior muscles

Photocopy this onto paper, or better still onto transparent plastic. Then use water-soluble pens to practise drawing in and labelling the muscles, until you know them without looking at the answers to help you!

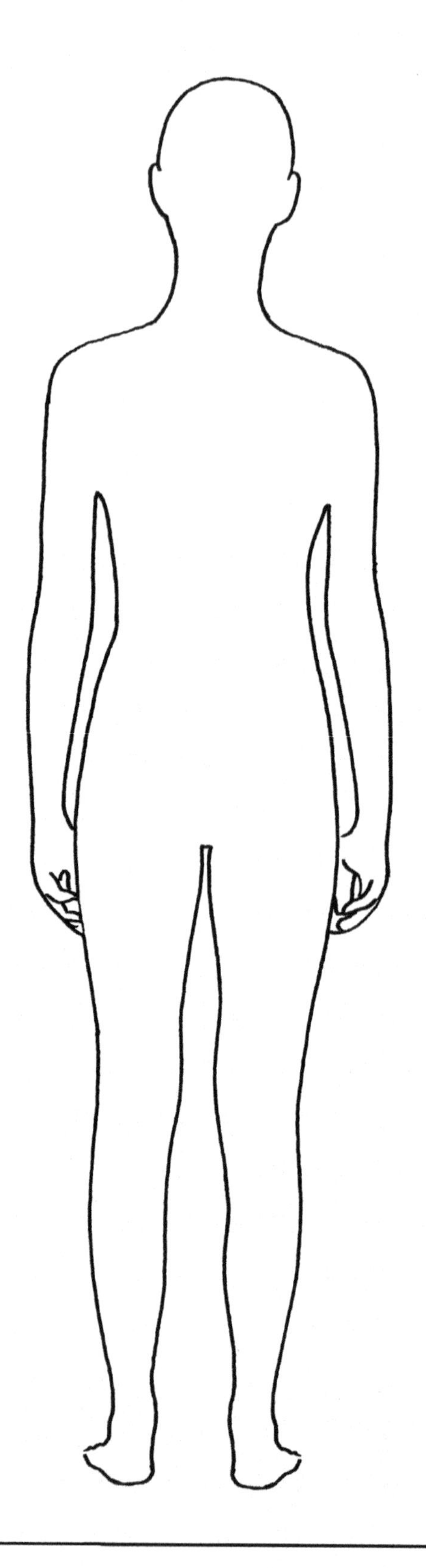

Template: anterior muscles

Photocopy this onto paper, or better still onto transparent plastic. Then use water-soluble pens to practise drawing in and labelling the muscles, until you know them without looking at the answers to help you!

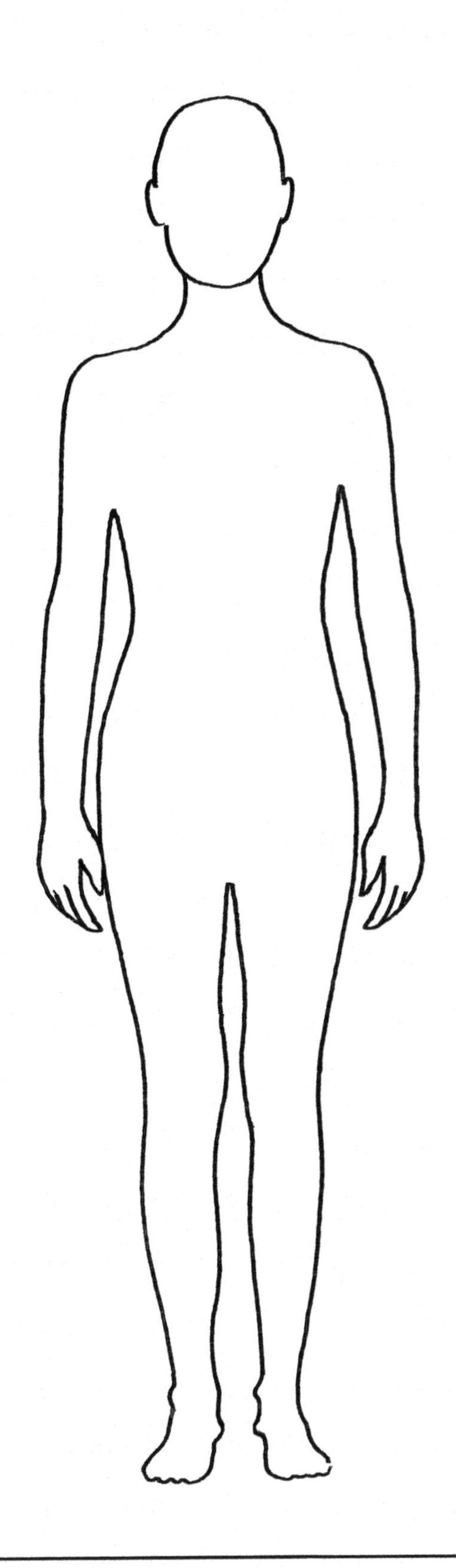

Activity: Colour-coding and labelling shoulder, arm and back muscles

The more different senses you use to remember facts, the more secure they will be in your memory. Here you will use hearing, movement and sight (colour, shape, layout etc.).

1 Study the diagram of the **shoulder, arm and back muscles**, repeating the names of the muscles aloud. See p. 91
2 Label and colour in the diagram below: one colour for each muscle.
3 Check your answers are correct. If not, start again straight away . . .

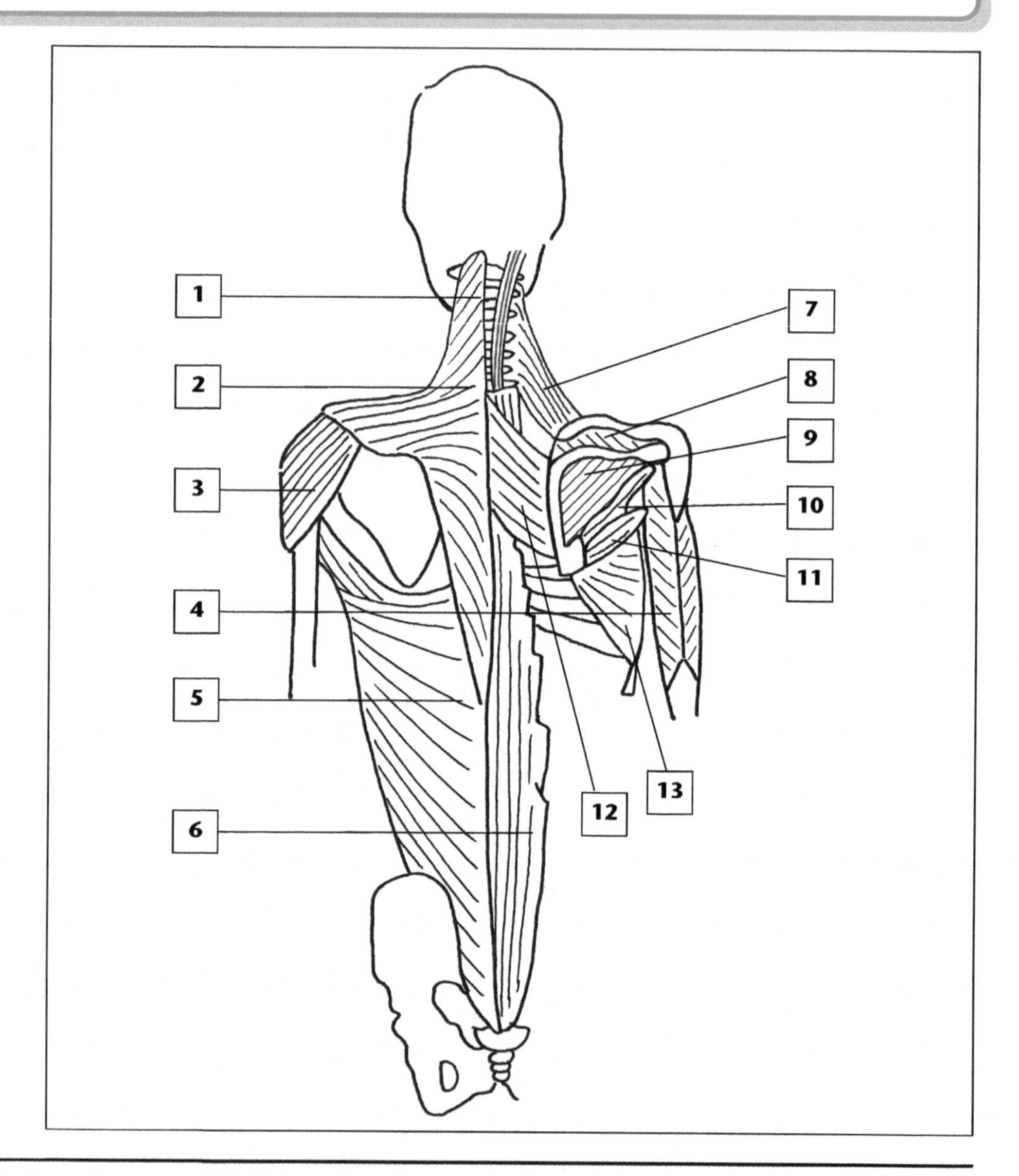

	Muscle	Function	Additional functions
1	**Splenius Capitis**	*extends* **head**	
2	**Trapezius**	*extends* **head**	*lifts, draws back* **scapula**
3	**Deltoid**	*extends* **arm**	*flexes, abducts* **arm**
4	**Triceps**	*extends* **elbow**	
5	**Latissimus Dorsi**	*extends* **arm**	*adducts, medially rotates* **arm**
6	**Erector Spinae**	*extends* **spinal column**	
7	**Levator Scapulae**	*lifts* **scapula**	
8	**Supraspinatus**	*abducts* **arm**	
9	**Infraspinatus**	*laterally rotates* **humerus**	
10	**Teres Minor**	*laterally rotates* **humerus**	
11	**Teres Major**	*medially rotates* **humerus**	
12	**Rhomboid**	*draws back* **scapula**	
13	**Serratus Anterior**	*draws forward* **scapula**	

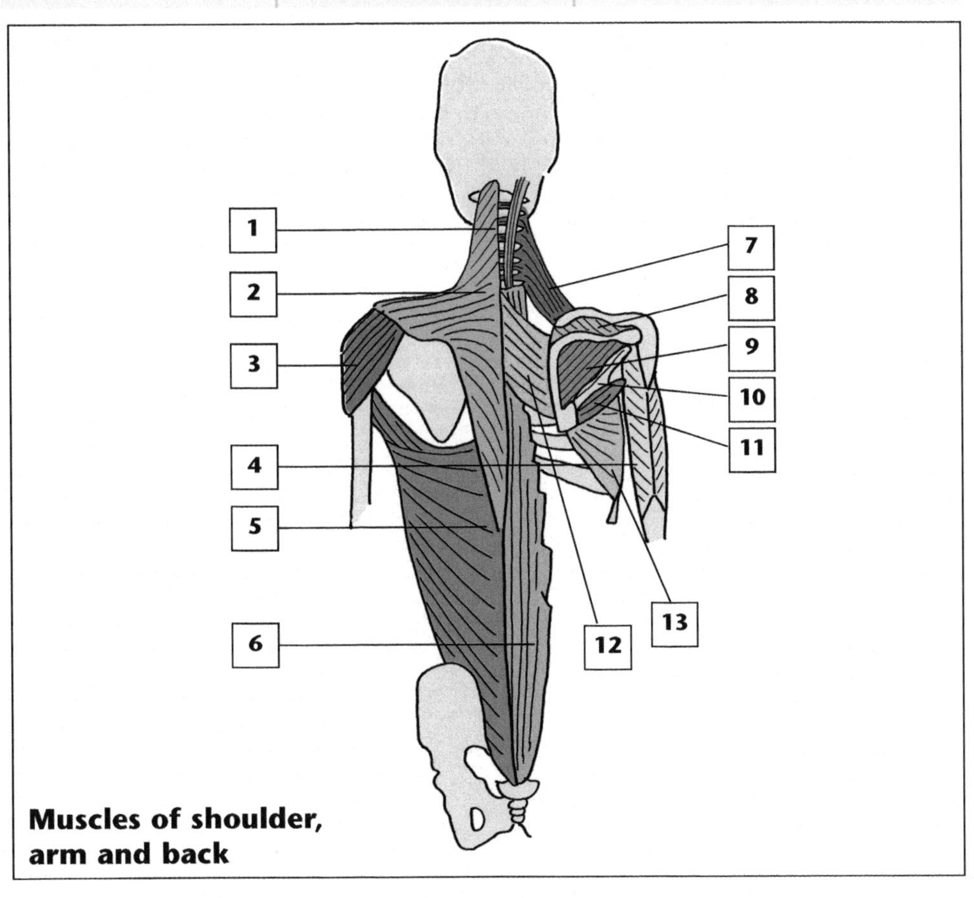

Muscles of shoulder, arm and back

Activity: Labelling lateral view of the brain

- Read the definitions below.
- Then label the blank diagram in pencil.
- Check your answers on the next page.
- Repeat until perfect!

Cerebellum: below the back of the cerebrum – it regulates balance, movement, posture, and muscle coordination.

Cingulate Gyrus: in the medial part of the brain – it functions as an integral part of the limbic system, which is involved with emotions, learning and memory.

Corpus Callosum: a large bundle of nerve fibres that connect the left and right sides (hemispheres) of the brain.

Medulla Oblongata: in the lowest section of the brainstem at the top of the spinal cord – it controls automatic functions such as breathing and heartbeat.

Occipital Lobe of the Cerebrum: at the back of each cerebral hemisphere – it contains the centres of vision and some reading processes.

Pituitary Gland: attached to the base of the brain – it secretes hormones.

Pons: just above the Medulla Oblongata – joins the hemispheres of the cerebellum and connects the cerebrum with the cerebellum.

Spinal Cord: a thick bundle of nerve fibres running from the base of the brain to the hip area, through the spine. It is the main pathway for information connecting the brain and the peripheral nervous system.

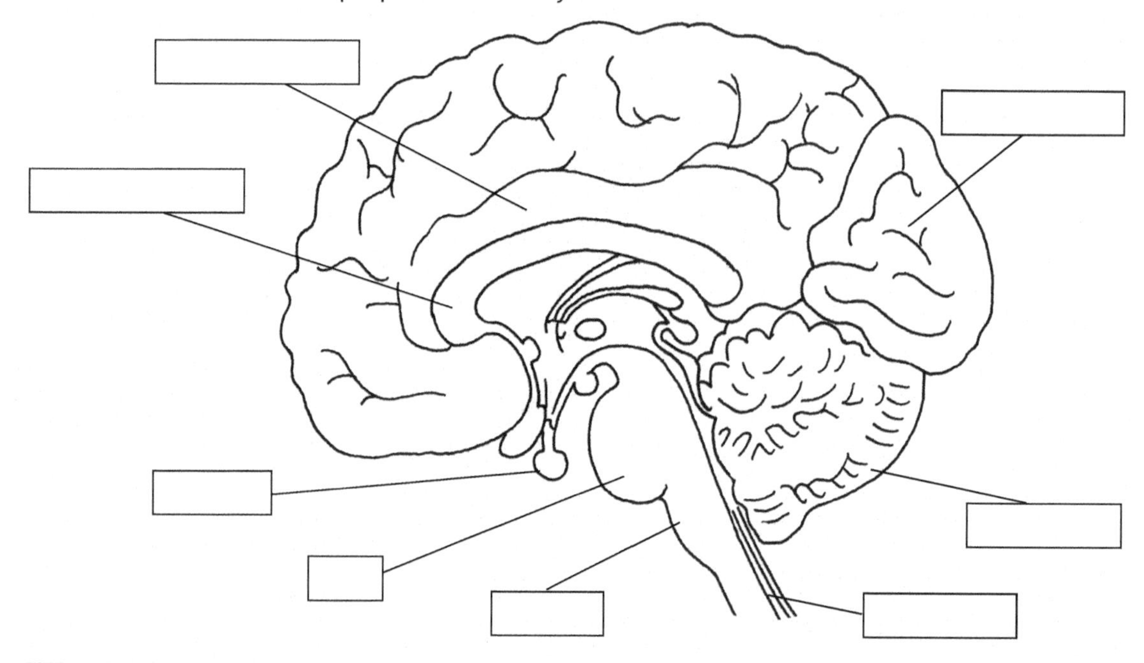

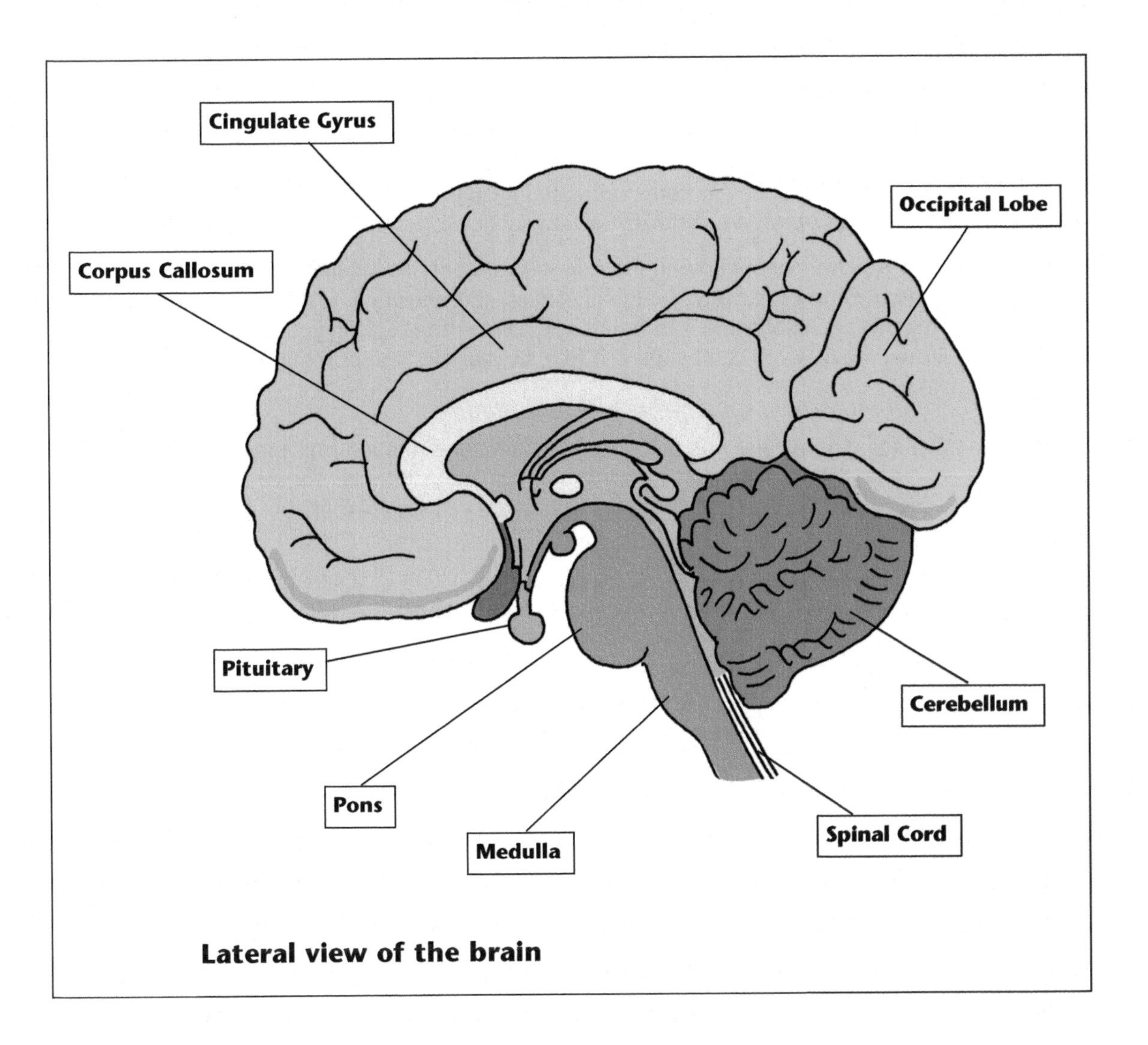

Lateral view of the brain

Did you successfully label this diagram? If so, why not use the crossword puzzle on page 129 to revise the Nervous System.

Posters

Uses of posters

An academic poster is usually a large piece of paper used to present a brief visual summary of a topic. It is commonly used for presenting research at academic conferences or as part of student assessments.

Here we also explore how to use a poster as a brilliant revision tool for learning Anatomy and Physiology. Designing a poster, reviewing it, discussing it with other students and explaining it to others will help fix the topic in your brain. As Albert Einstein said 'if you can't explain it to a six year old, you don't understand it yourself.'

What materials or equipment do you need?

Designing an A1 size poster for a professional presentation may present technical and communication challenges: you will need to use technology to produce a neat large scale poster. For the purposes of revision, you can simply use:

- coloured marker pens;
- a roll of plain wallpaper or flip chart paper;
- and cut and pasted printed images or headings.

Content

A poster has to convey a complex subject as simply and in the most eye-catching way possible. You shouldn't need extra notes or explanations to understand it.

1. Decide what is the main message or purpose of your poster.
2. Brainstorm, then identify the key points.
3. Establish the main title, headings and subheadings. Use only key words.
4. Add useful images or diagrams wherever possible.

Layout and visual presentation of your poster

It is important to consider how you might use arrangement, colour and writing to lead yourself or your reader through the content. The arrangement you choose for your poster depends on the task you have been given and the nature of your subject.

Presenting a piece of research

If you are presenting a piece of research, the structure of your poster will be similar to a formal report format:

1. Title
2. Introduction
3. Methods
4. Results
5. Discussion
6. Conclusion
7. References

Presenting a physiological process

Presenting a complex subject like a physiological process can be challenging as you need to reduce the information to its simplest visual form without dumbing it down. Physiology is the study of how cells, tissues and organs function: often cause-and-effect sequences are emphasized. This gives you a clue as to how this information can be visually represented – flow charts and bodymaps can be ideal. These are explored on page 96.

Presenting a memorable poster

If you want to be able to memorise the content of a poster, it is even more important to present the information in a clear, original and eye-catching manner. Below you can see an example of a poster from Public Health England, which makes use of a clever memory technique to help people remember how to recognize the signs of a stroke.

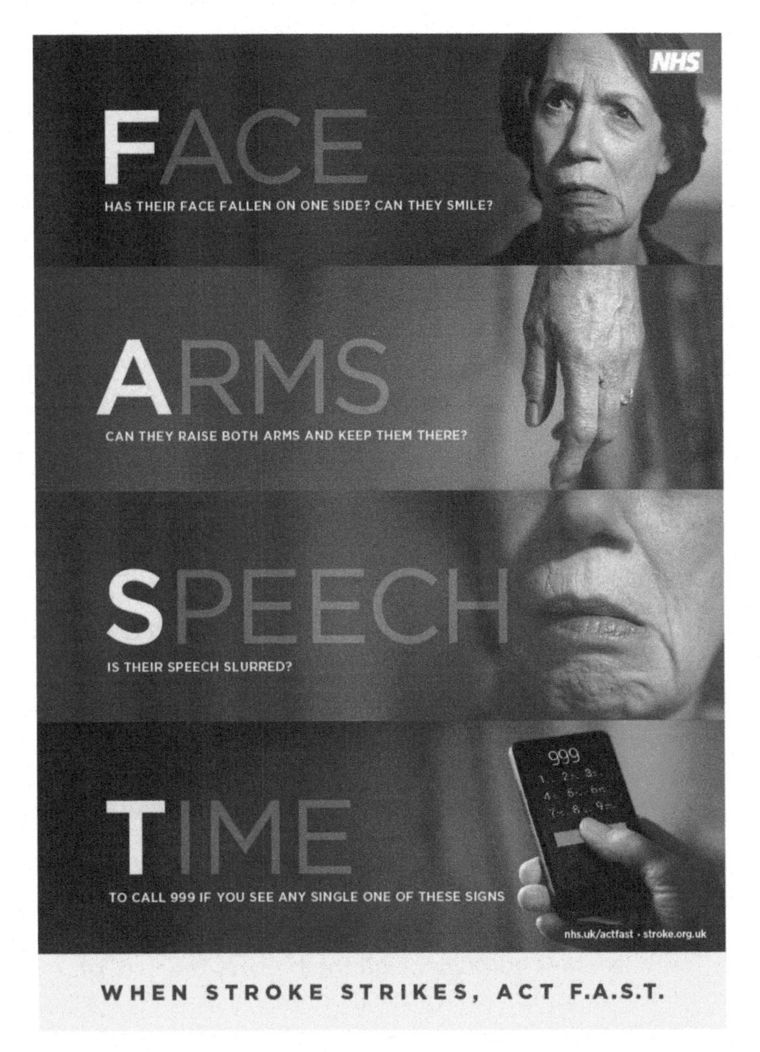

Useful software programmes

Mind Mapping software (see Useful Resources section on page 171) can help you brainstorm and experiment with different layouts before you make up your poster in a large A1 format.

In desk-top publishing or word-processing programmes, you will find many ready-made templates and SmartArt graphics that will inspire you to visually communicate your ideas. See some examples from Microsoft Word below:

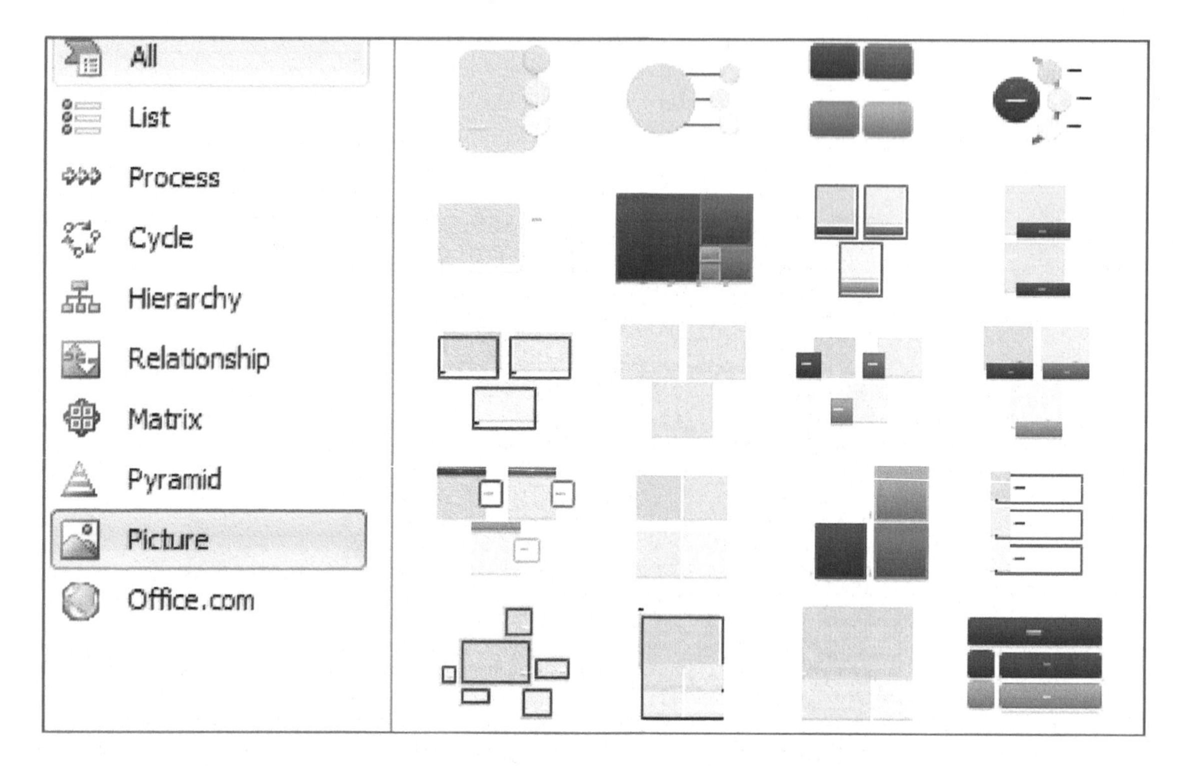

Here are some guidelines to help you maximize the visual impact of your poster:

- **Try to provide a clear visual entry point for readers, and a logical visual flow. Use numbers or arrows if content needs to be read in a particular order.**
- **Group related information: consider using boxes or lines round related information.**
- **Use margins and white space around your points to give your content room to breathe.**
- **If using a word processor to type and print text, use a plain sans-serif font like Arial.**
- **To be readable at a distance, the main title should be around 70 points, subheadings 40 points, and the body of the text 24 points.**
- **Every graphic should have a purpose and not just be there for decoration. Patterned backgrounds or random colours can be confusing and distracting.**

Memorising your poster

At a **formal poster presentation**, you would usually stand beside your poster, say a few introductory words as you point to headings or images, and then answer questions.

When **using a poster as a revision tool**, explaining the content to others is a very effective way to fix the information in your brain, especially as you are using so many different senses to store the memory: speech, vision and movement.

It is also important to **revise your poster several times** before your presentation or exam (see page 36 for guidance)

Summary: Using your poster for revision purposes

1. Is it immediately clear what the main topic and key points are?
2. Are the images and writing readable from about a metre away?
3. Have you used a maximum of about 300 to 500 words?
4. Can you represent anything visually in images or diagrams instead of using words?
5. Is it easy to understand what order your poster should be read in?
6. Have you tried presenting your poster to other students?
7. Have you got a plan and routine for revising this topic?

Activity: Draw a poster to illustrate the following

- How do you recognize the signs of a stroke?
- How does the endocrine system act to maintain homeostasis?
- How do neurons generate and conduct electrical impulses?
- Can you describe and illustrate the functions of the skin?

Then get together in a group and agree how to combine all your best ideas into one poster.

Chapter 6

Think-a-link

OVERVIEW

This chapter offers you opportunities to:

- ✓ **understand and explore the potential of the Link System of learning;**
- ✓ **practise this effective technique to remember names, processes and sequences of information.**

What is the Link System?

The Link System is based on centuries of usage and research. The Ancient Greeks had little access to writing materials, so they developed complex memory systems for remembering long stories, plays, poems and lectures. Countless researchers, students and even magicians have gone on to put this knowledge to good use for remembering items and sequences.

How does it work?

Most students find that the Link System is one of the easiest but most effective memory techniques to use. Basically you link a powerful image to an item on a list. Then you create a connection between it and the next item. When you learn to deliberately link items together with a memorable story featuring them, you will have a very powerful tool. You will be amazed how quickly you can train your memory to work more efficiently for you.

Why does it work?

The Link System uses the natural functions of your brain. Your memory operates by linking one piece of information with another. Think of how a particular smell or song can suddenly bring back a flood of memories.

Critics of this method would say that you can avoid understanding the complexities of the subject that is being studied and instead focus on memorizing the material so that it can be recalled parrot-fashion just for an exam. We would argue that this is the best way to learn some types of material – for example, the names of the synovial joints (see page 99) – and much better than mere rote repetition.

Naturally, when you have to learn a process such as the circulatory system (see 'The Blood Brothers' on page 102) or the functions of the kidneys (see 'The Kidney Kids' on page 103), it is very important to understand your subject fully in order to memorize it. The Link System can be used as extra reinforcement – a kind of insurance policy to make sure you can jog your overstretched memory during exams!

Activity: Using the Link System to learn the names of the joints

First read through the descriptions of the joints, below.

Test yourself half an hour later and note down how many you got right.

Now read the story on the next page and follow the simple instructions.

Test yourself again – you will be very surprised at the results!

- Ball and socket – e.g. hip and shoulder joints

- Hinge: for extension/flexion – e.g. knee and elbow

- Pivot: for rotation – e.g. radius and atlas

- Gliding – e.g. tarsal bone

- Saddle: for all movements except rotation – e.g. thumb

- Condyloid: for extension/flexion/adduction/abduction/circumlocution – e.g. wrist, first knuckle joint

Bend it like Granny Candy

My Gran is a sweet 80-year-old lady.
Her name is **Candy**.

Her latest boyfriend is called **Lloyd**.
(This will help you remember condyloid – get it?)

Anyway, she wanted to surprise **Lloyd** on his birthday with a special meal out.

Candy turned up at Lloyd's home on a 100 cc motorbike and made him climb up onto the **saddle** behind her.

She drove through the town at top speed.
When they finally arrived at the restaurant
Lloyd felt all un**hinged**.

Now things started to go from bad to worse.
Outside the restaurant some lads were playing football and Candy couldn't resist joining in.

Candy tackled a player and gracefully **pivot**ed round like a young David Beckham.

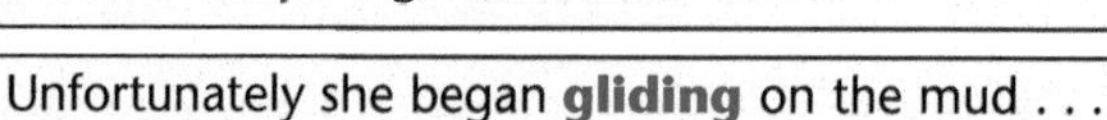

Unfortunately she began **gliding** on the mud . . .

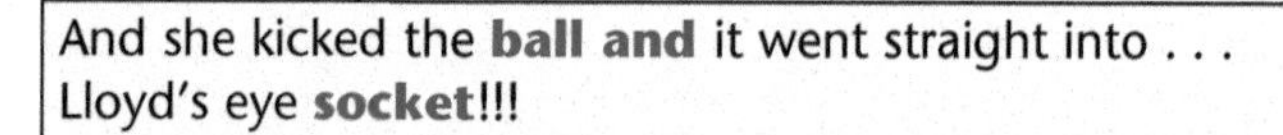

And she kicked the **ball and** it went straight into . . .
Lloyd's eye **socket**!!!

Candy drove Lloyd home at **6** o'clock with a black eye and his noise out of **joint**.

- Now read the story through again, stopping at each section to imagine the scenes and characters as fully as possible.
- Is Candy like your Gran or anybody else's you know?
- Imagine Lloyd bouncing up and down on the bike saddle and hanging on for his dear life.
- Feel the movements of Granny Candy pivoting, sliding, kicking the ball.
- Feel the shock of the ball in Lloyd's eye socket.
- Imagine Lloyd feeling sick and angry, unhinged and out of joint.
- Look at the clock and see a big red 6 (to remember the 6 kinds of joint).
- Play back the whole story in your mind whilst repeating the words aloud and stressing the ones in red.
- Tell somebody else the story, adding your own details. The sillier, ruder or more exaggerated the better.
- Now test yourself: I'm sure you won't forget the joints again and nor will the person you tell the story to. The Link System always works! Especially if *you* make up the story.

Remembering information in a sequence

Can you easily remember the circulatory system in the right sequence? And with all the right technical words? No? Well, you're just like the rest of us!

The best way to learn this kind of topic is to link all the bits of information into a story. This makes them instantly memorable. You can do this with almost *any* information, but you need to stretch your imagination. See, feel, hear, even smell the story in your mind! Make the images colourful, large, exaggerated, ridiculous or moving, like in cartoons.

- First, read over the whole story below, using the diagram on page 102.
- Read it over slowly, in small chunks, several times.
- Then, just using the diagram, try recalling the story from memory.
- Next try drawing the diagram from memory.
- Remember, if you want to fix this information in your memory, you will also need to revise it at specified intervals. See 'How can Bodymaps be used for exam revision?' in Chapter 2.
- Even better: make up your own story or diagram!

The Story of the Blood Brothers

Just imagine your body as a big factory with lots of different departments. Within the Circulatory Department, there are two faithful employees, nicknamed: 'The Blood Brothers'.

The Blood Brothers work day and night to bring nourishing blood (full of oxygen) round your body, and to remove used blood (full of waste products and carbon dioxide). This is the story of their journey round your body.

Big Ben Blood is the older brother: he is in charge of the **UPPER BODY**. Billy Blood is the younger brother: he is in charge of the **LOWER BODY**.

Big Ben collects all the used up blood from the Upper Body in a **blue** bucket and carries it (in a very superior way) down the **superior vena cava** into the **heart**. (To remember 'cava', imagine the vein is like a dark cave.)

Billy also collects the used up blood from the Lower Body in his **blue** bucket and walks (feeling rather inferior) up the **inferior vena cava** into the **heart**.

They both meet in the **right** entrance of the heart: the **right atrium**. Remember, they have to enter the heart in the 'right' way! They have to enter the heart by the **atrium** (a word for a 'posh' entrance).

Then a trap door opens: this is the **tricuspid valve**. With a rush they are sucked down into the right ventricle (imagine an air vent sucking them down). If you need to remember '**tricuspid**', imagine three cupids standing laughing at the trap door.

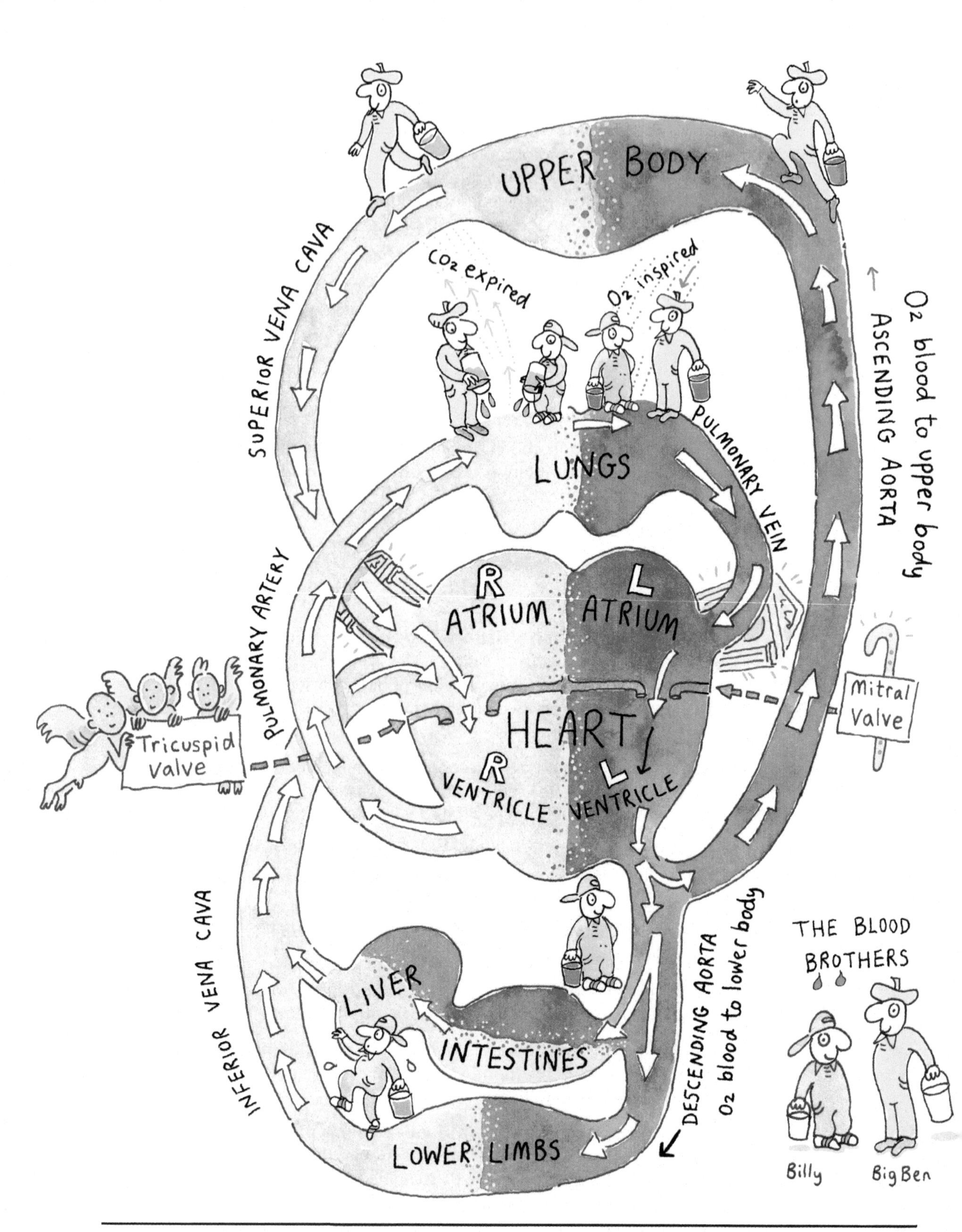
UPPER BODY
SUPERIOR VENA CAVA
CO_2 expired
O_2 inspired
O_2 blood to upper body
ASCENDING AORTA
LUNGS
PULMONARY VEIN
PULMONARY ARTERY
R
ATRIUM
L
ATRIUM
Mitral
Valve
Tricuspid
Valve
HEART
R
VENTRICLE
L
VENTRICLE
INFERIOR VENA CAVA
DESCENDING AORTA
O_2 blood to lower body
LIVER
INTESTINES
LOWER LIMBS
THE BLOOD
BROTHERS
Billy
Big Ben

Then the brothers leave the heart together. How? Imagine the two of them being squeezed up by a contracting muscle through a thick, smooth, **blue** tunnel called the **pulmonary artery** to the **lungs**.

In the lungs, **carbon dioxyde** (CO_2) is blown out of the blood in their **blue** buckets. They pour the blood into **red** buckets into which **oxygen** (O_2) is now sucked.

They now leave the lungs and travel through a thin **red** tunnel called the **pulmonary vein** back to the heart. They enter the heart by the posh **left atrium**.

Then a trap door opens (the **mitral valve**) and with a 'mighty' rush (yes, to remember the 'mitral' valve) they are sucked down into the **left ventricle**.

This is where they part company for a while: Big Ben ascends the **ascending aorta** to feed the upper body with clean oxygenated blood. Billy descends the **descending aorta** to feed the lower body with clean **oxygenated** blood.

And so continues their tireless journey round and round your body's circulatory system – that is, unless there's a breakdown somewhere in the system, but we won't talk about that sad state of affairs at this point . . . You've got enough to remember!

Activity: Using rhythm and rhyme

Rhythm and rhyme are powerful memory aids: this is why we can't get a silly advertising jingle out of our heads, even when we want to. When linked with vivid colourful images, as well as a storyline, they draw powerfully on your auditory and visual memory. Humorous, rude, silly or peculiar things tend to stick even more in your mind. The poem below can help you remember the function of the kidneys. Remember that your mnemonic will be more powerful if you use all your senses to code the information or dress up the images. When you re-read the poem, spend some time imagining and exaggerating the images, movements, sounds, smells, touch and emotions. Inventing your own rhymes can be even more powerful.

The Song of the Kidney Kids

Let's sing of the Kidney Kids, Kevin and Kate,
Who both do a job anybody would hate.
They collect up used blood both by night and by day,
Which they filter to take the waste products away.

Salts, drugs, uric acid, creatine and urea
Are removed with some water to leave the blood clear.
Then the toxic substance into urine is made,
As frothy and yellow as fresh lemonade.

This is poured through ureters – there are two of those –
Into the bladder and out through a hose
That is called the urethra. Your blood is now clean:
No one would ever know where it had been!

Chapter 7

Memory tricks

OVERVIEW

This chapter offers you opportunities to:

- ✓ **understand the potential of humour and mnemonics for learning;**
- ✓ **explore several different mnemonic systems;**
- ✓ **practise this effective technique to remember terminology, lists of words and numbers.**

What are mnemonics?

The word 'mnemonic' (pronounced 'ne-mo-nic') comes from 'Mnemosyne', the Greek Goddess of memory (she is usually depicted in art with a full mane of rich auburn hair). She was considered one of the most important and powerful goddesses of her time. After all, memory allows us to reason, to learn and to predict outcomes. It is the very foundation of language, culture and civilization.

The story goes that Mnemosyne gave some stunning gifts to human beings. She had responsibility for the naming of all objects, and therefore she gave us the means to speak to each other. Memory was of the greatest importance at the time of Mnemosyne, as vital information was mainly passed on via the spoken word.

How do mnemonics work?

Mnemomics are clever systems that you can use to remember information you are learning. Basically you link strange, silly or rude pictures and words to the information that you want to memorize. If you are worried that this might sound childish or ridiculous, remember that humour is one of the most effective tools for relieving tension and stimulating your brain. Look for humour in every learning situation and use it to your advantage.

Tricks for remembering words

For example, to remember that the **gracilis** muscle flexes your knee, imagine this picture of a flamingo trying to make a **graceful** curtsy to the audience.

To remember that the **serratus anterior** pulls your shoulder blades forwards, imagine using the **serrated** edge of a saw to cut something, and your shoulder blades pulling forwards with the motion.

Names can be tricky to remember. One of my lecturers was called Mr **Elsmart**: I used to imagine him with an elephant's head, wearing a smart suit.

To remember '**Gastrocnemious**', a muscle that crosses over the knee, first split the word into segments:

gas – troc – ne – mi – ous

Then visualize the following in great detail:

A gas (for **gas**) carrying truck (for **troc**) hitting me on my knee (for **ne**) near my (for **mi**) house (for **ous**).

Finally, say 'gas – truck – knee – my – hous' very fast several times! Sounds ridiculous? Maybe, but it works.

To remember the 7 functions of the skin

Memorize this sentence: '**A** **P**ale **S**kin **T**urns **C**rimson **V**ery **E**asily.'

A for '**A**bsorption'
P for '**P**rotection'
S for '**S**ense organ'
T for '**T**emperature'
C for '**C**ommunication'
V for '**V**itamin production'
E for '**E**xcretion'

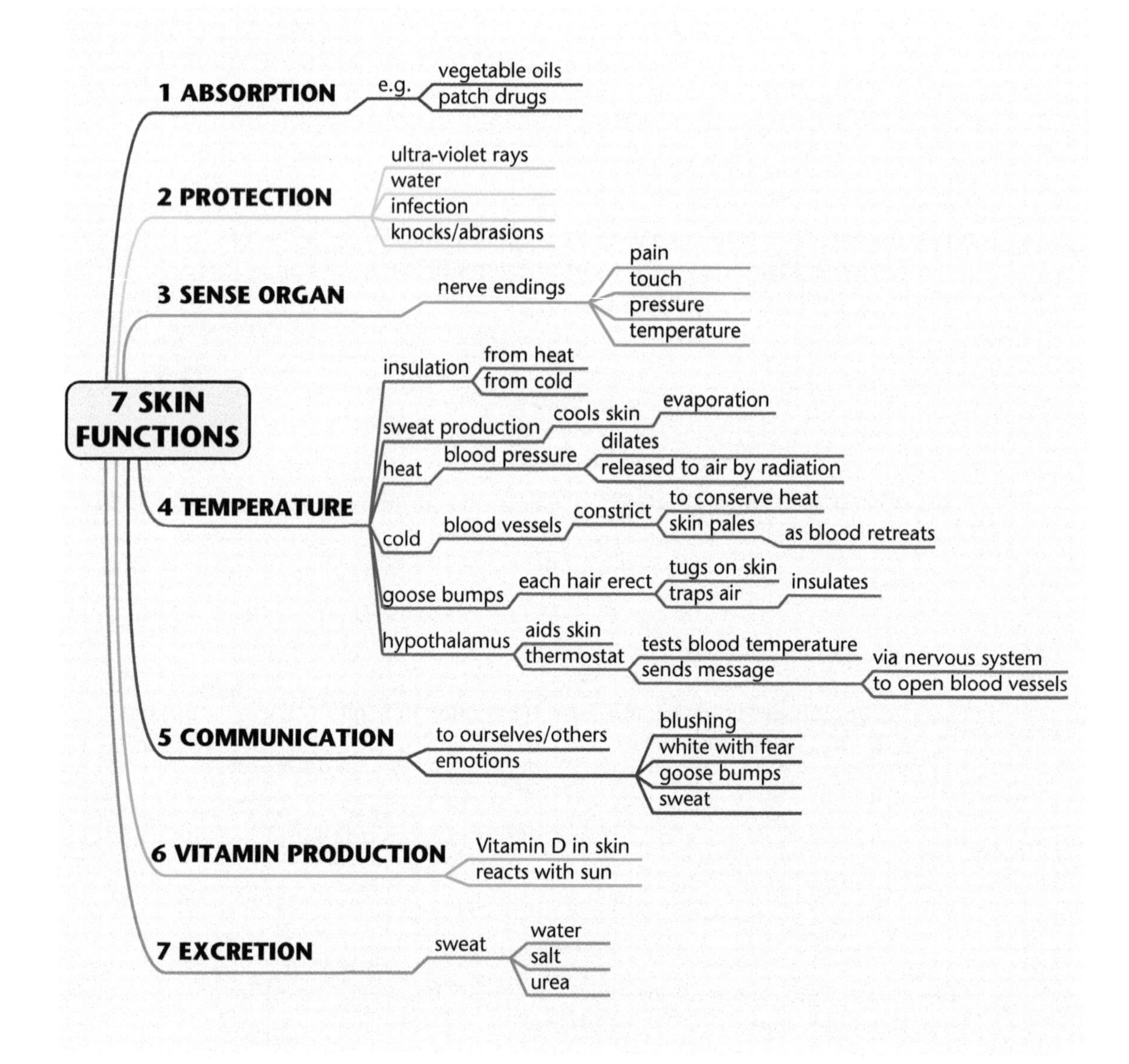

To remember the different types of muscle disorders

Memorize this sentence:
'**B**ruce **r**epeatedly **f**ibbed about **f**atigue, **s**train, **s**pasms and **f**rozen shoulders.'

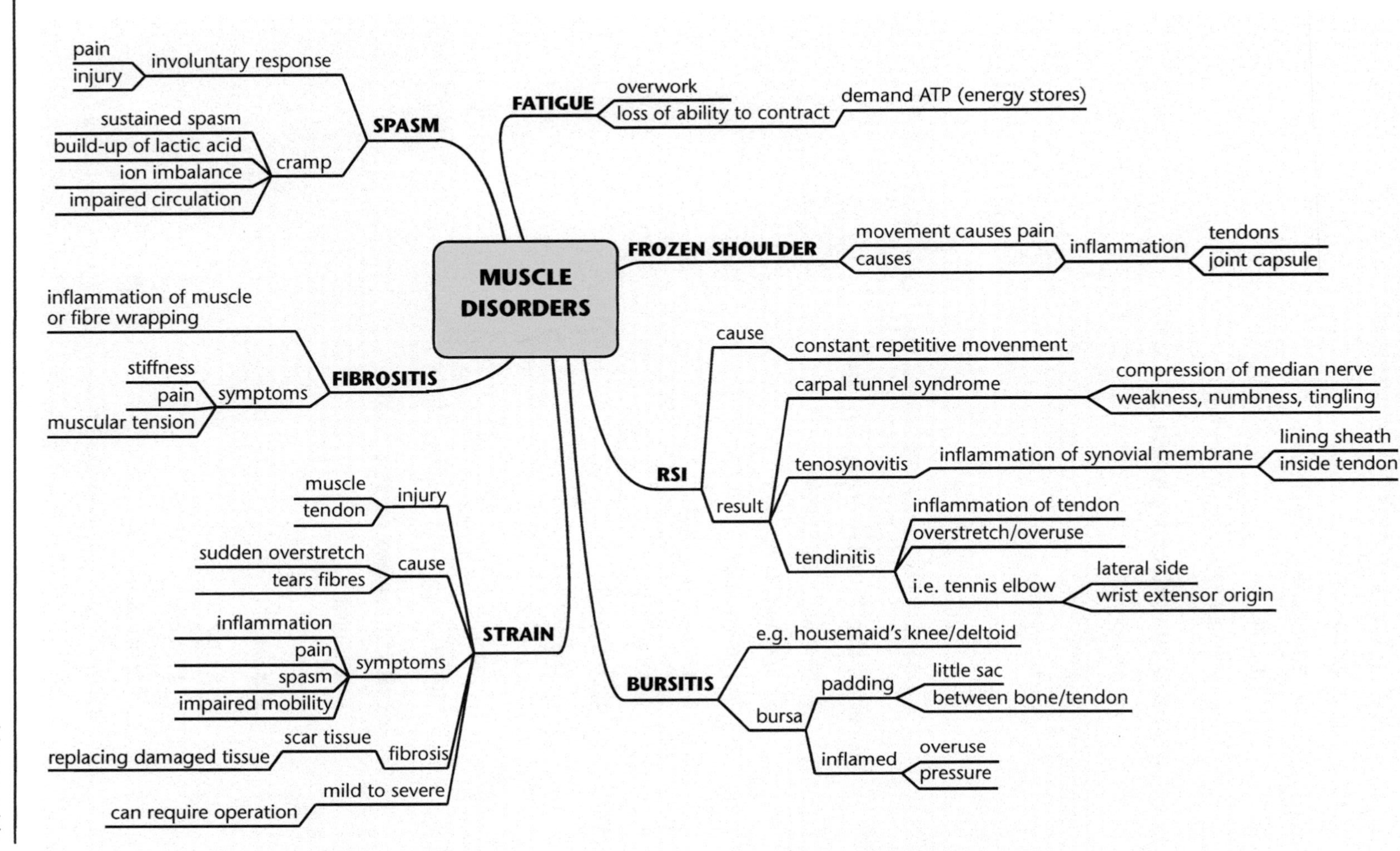

To remember the muscles of the feet

- **P**eroneus
- **E**xtensor digitorum longus
- **T**ibialis

Imagine a **PET** sitting at your feet

To recall the mitotic phases of a cell

- **I**nterphase
- **P**rophase
- **M**etaphase
- **A**naphase
- **T**elophase

Memorize this sentence: '**I** **P**roposed **M**arriage to **A**nna by **T**ext.'

To recall the 8 wrist (carpal) bones

Starting from the thumb:

- **S**caphoid
- **L**unate
- **T**riquetrum
- **P**isiform
- **T**rapezium
- **T**rapezoid
- **C**apitate
- **H**amate

Memorize this sentence: '**S**ensible **L**overs **T**ake **P**recautions **T**hey **T**hink **C**an **H**elp.'

To recall the bones of the skull

- **O**ccipital
- **P**arietal
- **F**rontal
- **T**emporal
- **E**thmoid
- **S**phenoid

Memorize this sentence: '**O**dd **P**eople **F**rom **T**yneside **E**at **S**piders.'

To remember the muscles of the abdomen

Imagine somebody doing sit-ups and the muscles in their abdomen beginning to **TIRE**.

Transversus abdominis
Internal abdominal oblique
Rectus abdominis
External abdominal oblique

To remember the difference between 'supine' and 'prone'

When you are **supine** you turn your palms up: imagine holding **up** a bowl of **soup**. When you are **prone**, you turn your palms **down**: imagine you are pouring out what is in your bowl down onto the ground.

To remember the difference between fibula and tibia:

TIBia is the **T**hick **I**nner **B**one. **F**ibu**L**a is **F**iner and **L**ateral

Tricks for remembering numbers

In A&P, it is very useful to be able to remember numbers and quantities. For example, how many different types of joints are there? How many different vertebrae in the spine? And what is the normal diastolic blood pressure?

Why is it so hard to remember numbers anyway? The reason is that they are usually meaningless, boring or confusing things to remember. This is why you need to find a trick to make each number meaningful, interesting and unforgettable! Try the strategies shown below. They really work.

To remember the 29 vertebrae

First, make up a memorable sentence, where each word has the same number of letters as each number you want to memorize. Completely confused? Let's take an example: when I want to remember that there are **7 cervical**, **12 thoracic**, **5 lumbar** and **5 sacral vertebrae**, this is what I do:

> I imagine a massage client called Carole. She is very thin and the vertebrae in her spine stick out. In my mind, I exaggerate the image of her vertebrae sticking out. I imagine my hand running over her lumpy spine.
>
> Then I repeat the sentence: 'Carole's terrifically lumpy spine'. Why? well . . .
>
> **Carole's** has **7** letters, starts with '**C**' and stands for **7 c**ervical vertebrae.
> **terrifically** has **12** letters, starts with '**t**' and stands for **12 t**horacic vertebrae.
> **lumpy** has **5** letters, starts with '**l**' and stands for **5 l**umbar vertebrae.
> **spine** has **5** letters, starts with '**s**' and stands for **5 s**acral vertebrae.

You may think this is a bit roundabout, but I bet you won't forget Carole's terrifically lumpy spine.

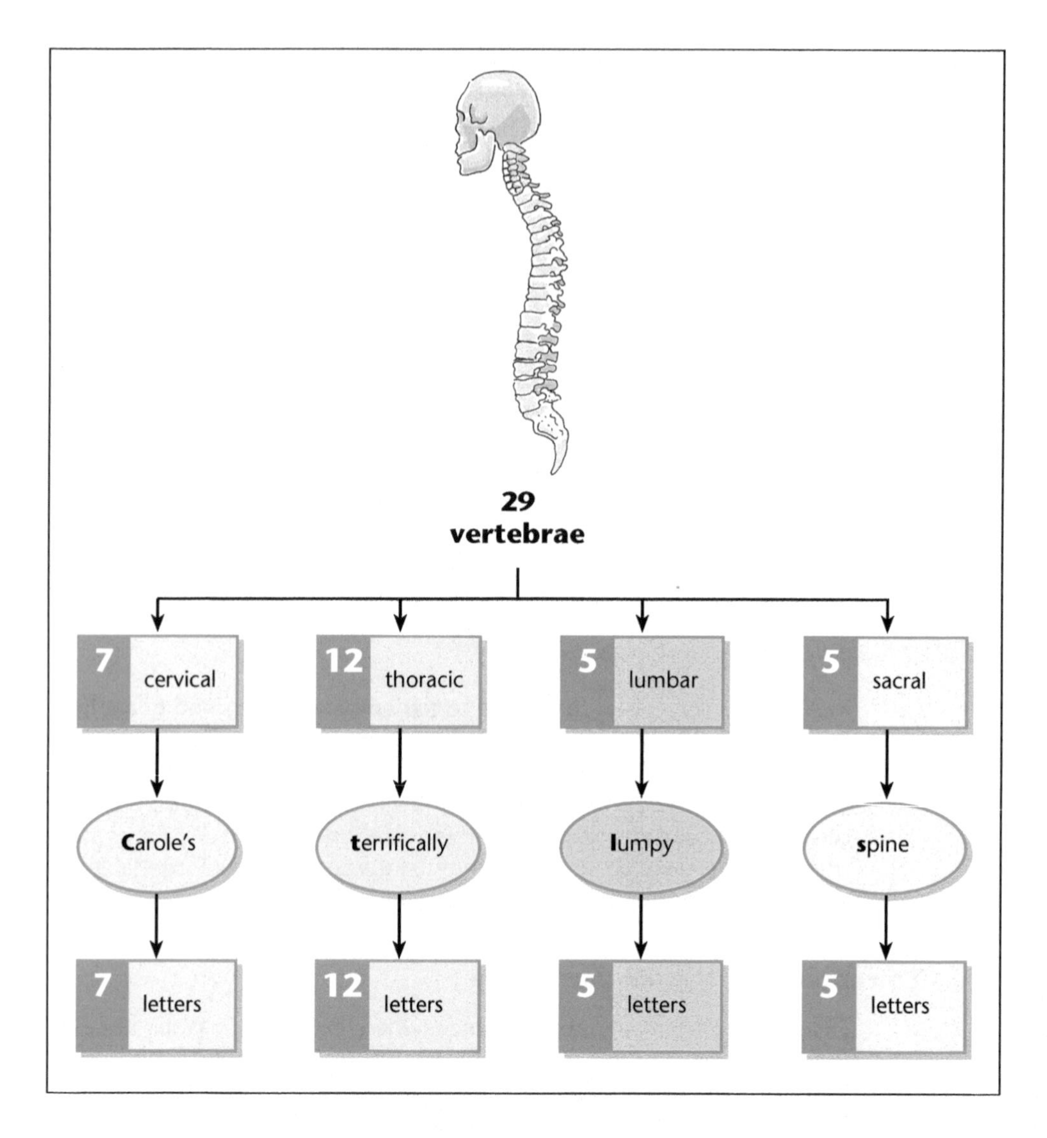

To remember a bank pin number

If you want to remember the number **4295**, for example, make up a sentence like this:

Grab	my	beautiful	money.
4 letters	**2** letters	**9** letters	**5** letters

To remember Celsius temperature

0°C . . . is ice.
10°C . . . is cool.
20°C . . . is nice.
30°C . . . need a dip in a pool.
37°C . . . is temperature of body.
100°C . . . would boil anybody!

Diagram created in Inspiration® by Inspiration Software®, Inc.

Remembering by imagining a journey

To remember that there are 11 different body systems

You can use familiar birthdays, ages, bus numbers, house numbers, telephone numbers and places as a strategy to remember unfamiliar numbers. The ancient Romans did this to remember long speeches that they had to make without using notes.

Try the following, it's worth the effort . . .

- Imagine getting on to a red number **11** tourist bus (take time seeing this number in big letters on the front of the bus in your mind's eye).
- In your mind, imagine the bus doing a round trip, visiting all the body systems one by one. You get off at each stop to visit the body systems and then get on the bus again.

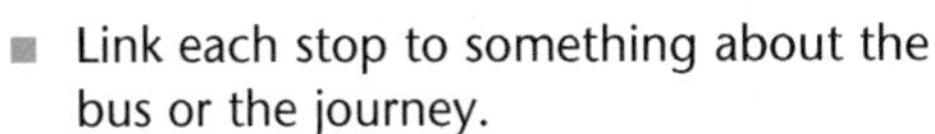

- Link each stop to something about the bus or the journey.
- The first stop is the heart and circulatory system. Link the red heart to the red bus and imagine your heart pounding as you climb to the top of a hill where the heart is. You will need to spend at least ten seconds imagining this scene, before you go on to the next stop.

Activity: Remembering the different body systems

Continue the bus journey in your mind's eye, stopping at all the body systems one by one. When you have finished, you will need to revise this journey several times in your mind, saying out loud the different body systems. This makes you use all your senses – such as movement, hearing and sight – at the same time. This is a very powerful memory strategy that you can apply to many topics in Anatomy and Physiology.

To learn the veins

This rather puffed-up looking guy is a member of a band called the 'Vein Men'. He can help you recall the **main veins in the body**. See illustration on page 113.

Activity: Remembering the main veins in the body

Get your favourite Anatomy and Physiology book out. Look up all the veins that you need to learn. Add them to the picture below. Or better still, make your own cartoon of a 'vein' person. It doesn't matter whether you think you can draw well or not. The act of drawing will help you remember the veins . . . And the sillier your drawing is, the more it will stick in your mind.

Add symbols to help you remember the different veins. For example, a **jug** full of wine for the **jugular** and an **axe** for the **axillary**.

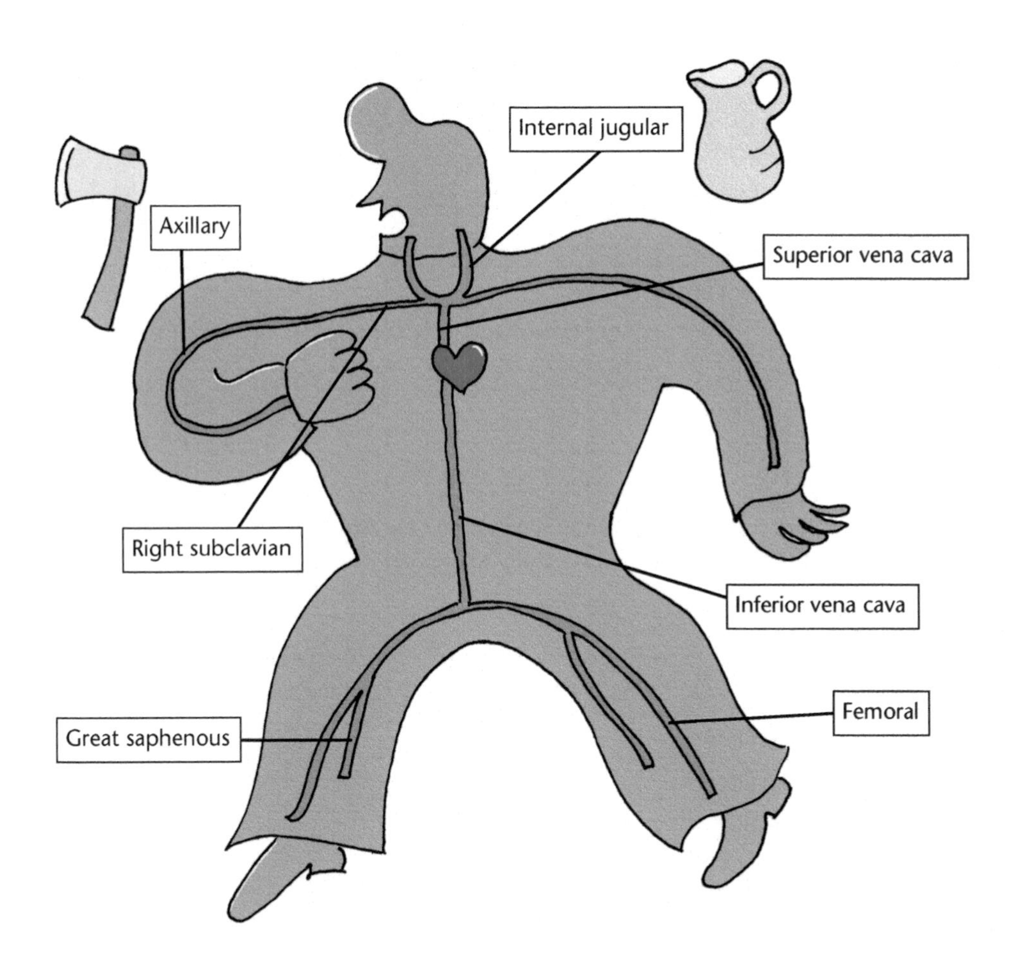

Chapter 8

Spelling: the 'WAM' way!

OVERVIEW

This chapter presents you with a solution for remembering difficult spellings.

- ✓ **You will learn how to use one of the most effective spelling systems for adults: the 'WAM' Spelling System.**
- ✓ **You will practise terminology relating to the Body Systems using word games and crosswords.**

What does '*WAM*' mean?

'WAM' stands for: **W**rite ⟶ **A**nalyse ⟶ **M**emorize.
Don't worry: the method is fully explained in a moment.

When is spelling important?

So many words in English are quite horrible to spell, especially those from Anatomy and Physiology! Some are very difficult to pronounce as well – what about:

- *Langerhans*?
- *aponeurosis*?
- *brachioradialis*?

If you can't remember how to spell a word, it can:

- interrupt your thinking;
- make you lose precious time;
- give a poor impression or confuse your reader; and
- give your confidence a knock.

Why is WAM different from other spelling methods?

- You won't be working through pages of confusing spelling rules.
- What's more, you won't be copying out words a 100 times.
- There will be no humiliating tests.
- You won't need to change the way you speak.
- You will not be told that to read more will improve your spelling: that's just not true!
- You certainly won't be told that you are 'lazy, thick, careless or hopeless'.

WAM has helped many thousands of adults solve their spelling difficulties. There are 3 main steps in the WAM method. These are described over the next pages.

The 3 steps

If you want WAM to work, you will need to really follow the 3 steps and explanations below . . .

STEP 1 Write

From your own writing, carefully choose **7 words** that are giving you trouble to spell. These must be **important** words for you to know.

Do not learn more than 7 new spellings in one week, as you can overload your memory and confuse yourself. It's better to create very strong memory traces for 7 words than weak traces for 20 words!

As the weeks go by, you will be amazed. You will build up a large stock of common spelling patterns, as well as useful words. You'll soon be able to spell hundreds of new words with the same spelling patterns in them.

STEP 2 Analyse

This step involves you **examining each word** from your list of 7 words very carefully.

Take each word, one by one. Imagine putting it under a microscope. Now dissect it! Analyse carefully which bit or bits are giving you most trouble. Use a highlighter pen to highlight the hard bits.

For example: lymphatic
epicraneas

STEP 3 Memorize

There are **6 different memory techniques** for you to try out. For each word, there will be the perfect technique. All you have to do is experiment with each one and see what works best for you.

The 6 different memory techniques are demonstrated on the next pages.

The 6 different memory techniques

1. Cut-it-up

Cut the word into smaller bits.
If possible, try to find small words within larger ones.

For example:

separate = **sep / a / rat / e**
langerhans = **l / anger / hans**
epicraneas = **epi / crane / as**
brachioradialis = **bra / chi / ora / dial / is**

2. Say-it-your-way

Change the way you say the word, so that you can hear every letter. Tap the rhythm out.

For example:

Wednesday = '**wed**' - '**nes**' - '**day**'
cleido = '**clee**' - '**i**' - '**do**'

3. Think-a-link

This is a way of linking a word you don't know to a word you **do** know. For example:

- Think of a visual link:

  neurotic
  aponeurosis
  pectineus

- Think of a link by sound – these words all rhyme:

  enough
  rough
  tough

- Think of a link by meaning:

  hyper means 'excessive' 'over the top'
  hypersensitive means 'too sensitive'

See page 122 for tips on how to tackle terminology.

4. Change-the-look

- Colour-code the word: **sphenoid**

- Illustrate it: sp**hen**oid

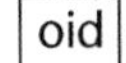

- Write it, cut it up and rearrange the bits

| oid | | sp | | hen | = | sp | hen | oid |

5. Memory tricks

These are often called 'mnemonics': they are simple tricks to help you remember something.

For example, to remember 'diarrhoea':
Make up words into a silly sentence using all the letters in diarrhoea . . .

diarrhoea
is
a
really
really
horrible
odorous
expelling
activity

To remember 'rhythm':
rhythm **h**as **y**our **t**wo **h**ips **m**oving

6. Spelling rules!

Has anybody ever told you that, if you learn all the rules of English spelling, your spelling problems will vanish? The truth is that very few rules in English are really useful. First, there are loads of exceptions to these rules. Secondly, some rules are far too complicated to be useful. Having said all that, here are 6 rules that come in very useful.

i' before 'e' rule

Put 'i' before 'e'	believe
except after 'c'	receive
Exceptions:	neighbour
	weight
	height

Very important:

- **Do not learn confusing words like 'believe' and 'receive' in the same week.**
- **Keep them as far apart as possible, or they will fight against each other in your memory.**
- **Learn one pattern first. When you are completely confident with that one, you can introduce a second pattern.**

Plurals

Plural means 'more than one'.
For example: 'bones' is the plural of 'bone'.

- Most words are made plural by adding 's'.

For example:	one cell	two cells

- For words ending in 's', 'ss', 'ch', 'sh', 'x', you add 'es'

For example:	one bus	two buses
	one cross	three crosses
	one wish	many wishes
	one box	some boxes

- For words ending in 'f' change the 'f' into a 'v' and add 'es'

For example:	one leaf	many leaves

There are some words that come from Latin and Greek and do not follow the above rules (for example: 'neuroses', 'papillae' and 'nuclei' are all plurals!). They must be learnt one by one, using a different memory technique.

Double-Trouble

Every word is made up of one or more **syllables**.
A syllable is a part of a word: it is made by one push of breath.
There are 4 syllables in the word 'television': **te** - **le** - **vi** - **sion**.
When a syllable is **stressed**, it means you say that bit with more weight on it.

A **vowel** is a sound made by your vocal cords when they vibrate.

There are **5 vowels** in the alphabet: **a**, **e**, **i**, **o**, **u**.
'**y**' is a vowel when it makes an 'ee' sound.

Every vowel can have a **long** or a **short** sound, but it depends where it appears in a word, for example:

Listen to 'a' in hat or hate
Listen to 'e' in pet or Pete
Listen to 'o' in rod or rode
Listen to 'i' in pin or pine

Every letter that isn't a vowel is a **consonant**.
The consonants are: **b**, **c**, **d**, **f**, **g**, **h**, **j**, **k**, **l**, **m**, **n**, **p**, **q**, **r**, **s**, **t**, **v**, **w**, **x**, **z**.

- In short words of one syllable, with one short vowel and ending in a consonant:

  double the final consonant before adding an ending such as 'ed' 'er' 'est' 'ing'.

  For example:

	run	+ **ing**	=	**running**
	fit	+ **er**	=	**fitter**
	drop	+ **ed**	=	**dropped**
	big	+ **est**	=	**biggest**

- In words of two syllables, if the stress is on the second syllable:

  double the consonant before adding the ending.

  For example: **admit** + **ed** = **admitted**

- In words ending in 'l':

  double the 'l' before adding the ending.

  For example: **travel** + **ed** = **travelled**

Silent 'e'

- Drop the final silent '**e**' when adding endings beginning with a vowel such as: '**ed**', '**er**', '**est**', '**ous**', '**ing**'.

For example:

hate + **ing** = **hating**
fame + **ous** = **famous**

- But, for words ending in '**ce**' or '**ge**' keep the '**e**' when adding '**ous**' or '**able**'.

For example:

advantage + **ous** = **advantageous**
notice + **able** = **noticeable**

'y' at the end of a word

- The '**ee**' sound at the end of a word is written with a '**y**'.

For example: **funny**, **handy**

- When you add an ending such as: '**ed**', '**er**', '**est**', '**ment**', '**ly**', '**able**' to words ending in '**y**', the '**y**' changes to an '**i**'

For example:

rely + **ed** = **relied**
rely + **able** = **reliable**

'full' at the end of a word

When adding '**full**' at the end of a word, drop one '**l**'.

For example:

care + **full** = **careful**

Putting it all together

Your Weekly WAM Routine – this is important!

1. **Each week, make out a new list of seven words.**
Use WAM at least once a day for that list.
It should only take a few minutes, but it is really important to repeat the procedure regularly. It's like taking a pill: there's no point just taking it when you fancy it or when you remember!

2. **To test yourself: take each word, one by one.**
   - Bring back to mind the memory technique you used for that word.
   Picture the word in your mind.
   Say the word out loud, as you write it from memory.
   Use joined up handwriting: this will help you remember the flow and feel of the word as you write it.
   - Now check your version carefully against the original.
     - If it is correct, congratulate yourself.
     - If it is wrong, forgive yourself! Then go back to steps 2 and 3 straightaway.

3. **At the end of the week:**
   - If you really know the word, store it in a personal spelling notebook for future reference. A small alphabetical address book is perfect. (When you have a spare moment – in the bath or train or before going to sleep – you can always go through your spelling notebook.)
   - If you are still having problems with a particular word, put it back on the following week's list. This way, it will never come off your list until you know it. This is a very thorough and deep way of learning spelling.

Terminology: strange words used in Anatomy and Physiology

Most words in A&P seem to be written in some secret code. Some come from the name of a person. For example, the **Eustachian tubes** were named by an Italian anatomist called **Eustachi**. The first vertebra of your spine is called the **atlas**. In Greek mythology, **Atlas** was a very strong man who held up the world on his back. This gives us the picture of the atlas vertebra holding up the very heavy head.

In fact, most words in A&P are foreign as they come from Latin or Ancient Greek. This gives us some very long and strange looking words! But these can be tackled more easily when broken up into parts. Each part means something and when the parts are put together again, you will have broken the secret code! **If you can link these words to their original meaning and a picture, you will be able to remember them much more easily**.

Most words are made up of 2 parts:

- a **prefix**: at the beginning of a word;
- the **root**: the middle of the word and the most important part.

You probably already know many prefixes without realizing it: many common words in English make use of them: for example, the prefix '**tele**' means 'far away'. So we get words like:

- **Television**: to **see pictures** coming from **far away**
  Telephone: to **hear words** coming from **far away**
- **Gynaecology** means the study of the functioning of, and disorders affecting, the female reproductive organs.
  Gynae means **female** and **ology** means the **study of**
- **Psychology** means the study of the mind.
  Psyche means **mind,** and **ology** means the **study of**

In some cases, just one letter's difference in the prefix of a word can change the meaning totally. For example:

- **Abduction**: from '**ab**' meaning **away** and '**duction**' meaning **movement**
- **Adduction**: from '**ad**' meaning **towards** and '**duction**' meaning **movement**
  So **abduction** means moving a limb away from the side of the median line of the body.

There are many medical dictionaries – free online, available as computer programs or hand-held ones. See 'IT solutions for saving time' on pages 55–6.

Activity: Linking words

1 Can you link a word from the list below to a picture? If you are stuck use a dictionary of word origins to help you.

Cochlea **Dendrite** **Hippocampus** **Pineal** **Pons** **Insula**

2 Can you link a word from the column on the left to a definition on the right? There is a list of prefixes, roots and suffixes to help you.

Hydrophilic:	involving the heart and the blood vessels
Haemophiliac:	last phase of cell division
Bicuspid point:	cell movement
Erythrocyte:	first stage of cell division
Cytokinesis:	double pointed
Cardiovascular:	water loving
Endoscopy:	red blood cell
Telophase:	blood disorder due to deficiency of coagulation factor
Prophase:	internal examination

Prefixes, roots and suffixes

bi: two
cardio: heart
cuspid: pointed
cyte: cell
endo: inside
erythro: red
haemo: blood
hydro: water
kinesis: movement
phasis: appearance
philia: love of
pro: first, before
scopy: look
telo: last
vascular: vessel

The answers are given on page 172.

Crossword puzzles

Regularly doing word games is an enjoyable and effective way to help you revise and fix information – such as spellings and definitions that you need to memorize. The simple habit of doing crosswords has educational as well as health benefits:

- It improves general brain functions such as problem-solving, working memory and 'tip-of-the-tongue' word-finding.
- It causes the hormone serotonin to be released into the body, enabling the body and mind to better cope with stress.
- Just as physical activity strengthens the heart, muscles and bones, studies have shown that exercising the mind strengthens the brain against disease.
- The brain normally uses an enormous amount of the body's energy – about 20 per cent of your body's entire energy production. When you work your brain harder, the blood flow goes to the brain and you use more energy. The effect of brain-challenging activities seems to be to build up a reserve of neuron connections, stopping the richness of the connections between cells beginning to decline.
- Teaching your brain *new* tricks – such as revising in a *non-routine* way – stimulates new neural connections within the brain, leading to greater flexibility and boosting brainpower.
- Working together with other students on a crossword encourages discussion of meaning and spelling: this will help fix the words in your memory.

In this section, you will find a selection of crossword puzzles to help you revise spellings and specialist vocabulary. Fill in your answers in pencil so that you can rub them out and do them several times in the period leading up to exams. You may also like to work in pairs or a group, as talking over clues will fix them in your auditory memory as well as your visual and motor memory.

The circulatory system crossword

(answers on page 172–3)

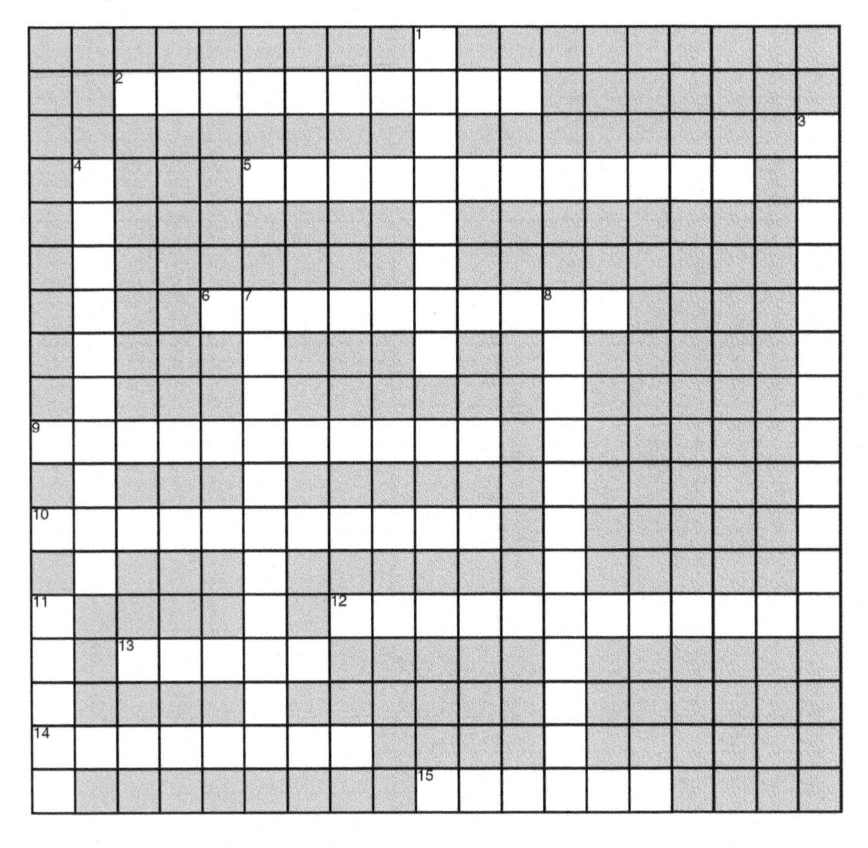

Clues Across

2. This is the middle layer of the muscular wall of the heart and is made of muscular tissue.
5. These make up 75% of the leucocytes and defend the body against viruses and bacteria.
6. These are the bottom left and right chambers of the heart.
9. These are formed in lymphatic tissue and produce antibodies.
10. This covers the outside of the heart.
12. The process by which bacteria are devoured.
13. These are the top right and left chambers of the heart.
14. This is the name of the circulation from the heart to the body.
15. This makes up 55% of the blood and helps transport essential substances round the body.

Clues Down

1. Blood is pushed by contraction of the left atrium into the left ventricle through this valve.
3. These are a type of blood cell that help with blood clotting.
4. These protect the body from infection and increase by mitosis when the body is attacked.
7. This is the inner layer of the heart's muscular wall.
8. These are also known as red blood cells.
11. The rate at which your heart pumps blood through the circulatory system.

The digestive system crossword

(answers on page 173)

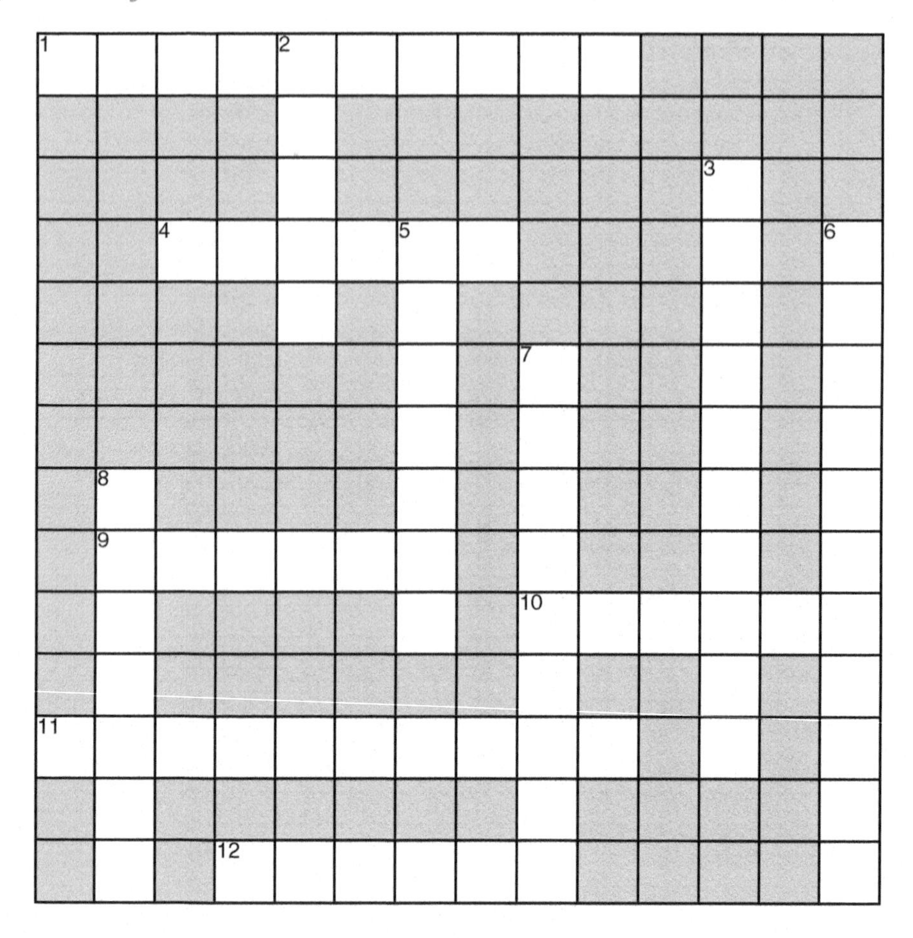

Clues Across

1. This prevents choking.
4. The unpleasant unwanted products of food.
9. This contains taste buds.
10. This is secreted by glands and helps start digestion.
11. These juices help digest food in the small intestine.
12. Enzyme that turns proteins into amino acids.

Clues Down

2. Sugars and amino acids are passed to this organ.
3. This is when digested food is absorbed through the walls of the villi.
5. These make food more digestible.
6. This type of movement pushes food against the villi.
7. Breakdown and transformation of food into substances transported by the blood.
8. Place where protein is digested.

The endocrine system crossword

(answers on page 173–4)

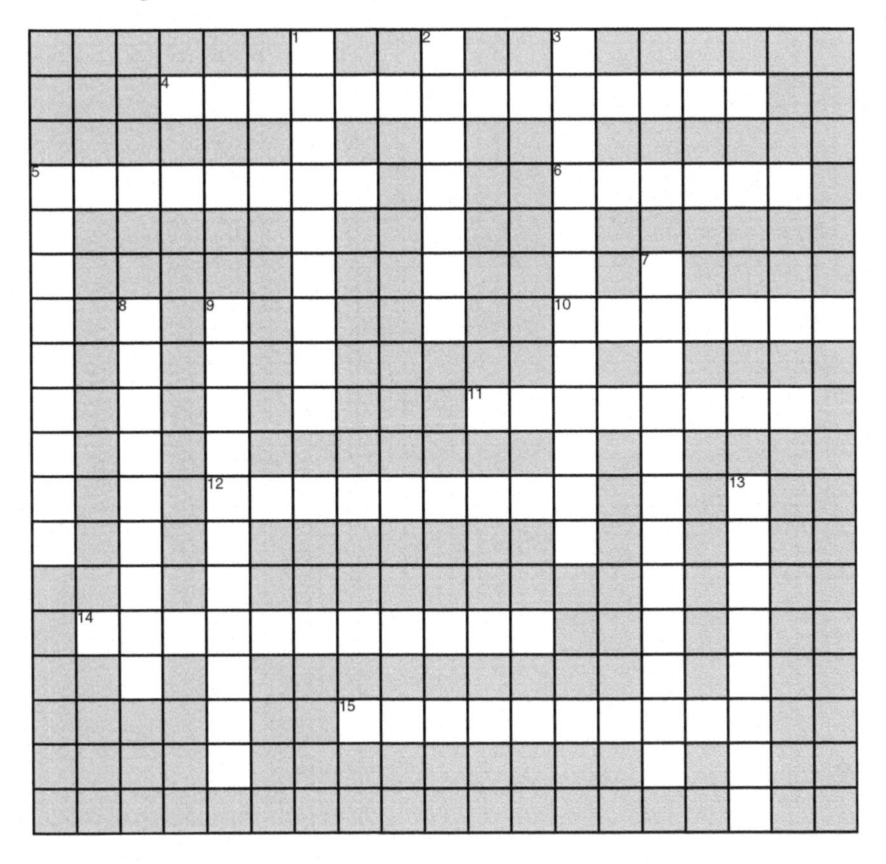

Clues Across

4. This means that too much of a hormone is produced.
5. This organ secretes insulin in the islets of Langerhans.
6. This gland is part of the immune system.
10. Hormones secreted by these glands stimulate growth.
11. These are chemical messengers produced by glands.
12. The ovaries secrete this as well as progesterone.
14. This kind of gland is situated on both sides of the thyroid gland.
15. This hormone is released to help us fight a threat or run away from it.

Clues Down

1. The pineal gland secretes this.
2. The inner part of the adrenal gland.
3. The testes secrete this.
5. This kind of gland is situated at the base of the brain.
7. This occurs when there is a lower than normal level of blood glucose.
8. This kind of rhythm means that regular fluctuations of hormone levels occur every 24 hours.
9. This means that the internal environment of the body is kept in balance.
13. A kind of gland which has no separate tube or canal to carry hormones into the bloodstream.

The lymphatic system crossword

(answers on page 174)

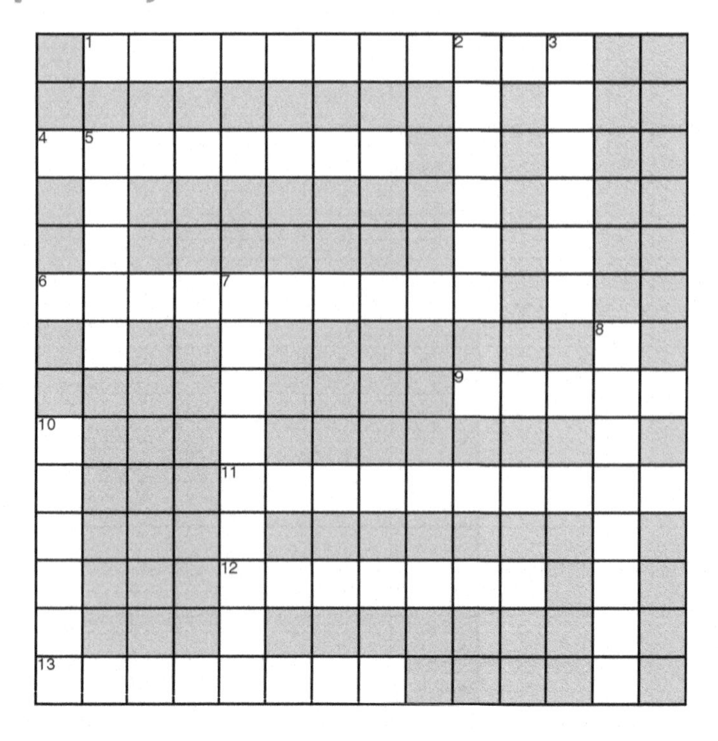

Clues Across

1. Lymph contains waste materials such as these.
4. These nodes are situated near the genital area.
6. Lymph contains waste materials such as these.
9. This fluid is similar to blood plasma and is filtered by the lymphatic nodes.
11. Lymphatic tissue contains cells such as these.
12. These contain lymphatic tissue and are situated in the throat.
13. These nodes are situated near the armpits.

Clues Down

2. This gland contains lymphatic tissue and is situated behind the sternum.
3. This is a non-essential organ which produces and destroys cells.
5. The lymph vessels open into these structures, which can be found throughout the body.
7. These nodes are situated at the back of the skull.
8. This non-essential organ contains lymphatic tissue and is part of the digestive system.
10. This is the swelling of tissues, caused by an obstruction to the lymphatic flow.

The nervous system crossword

(answers on page 175)

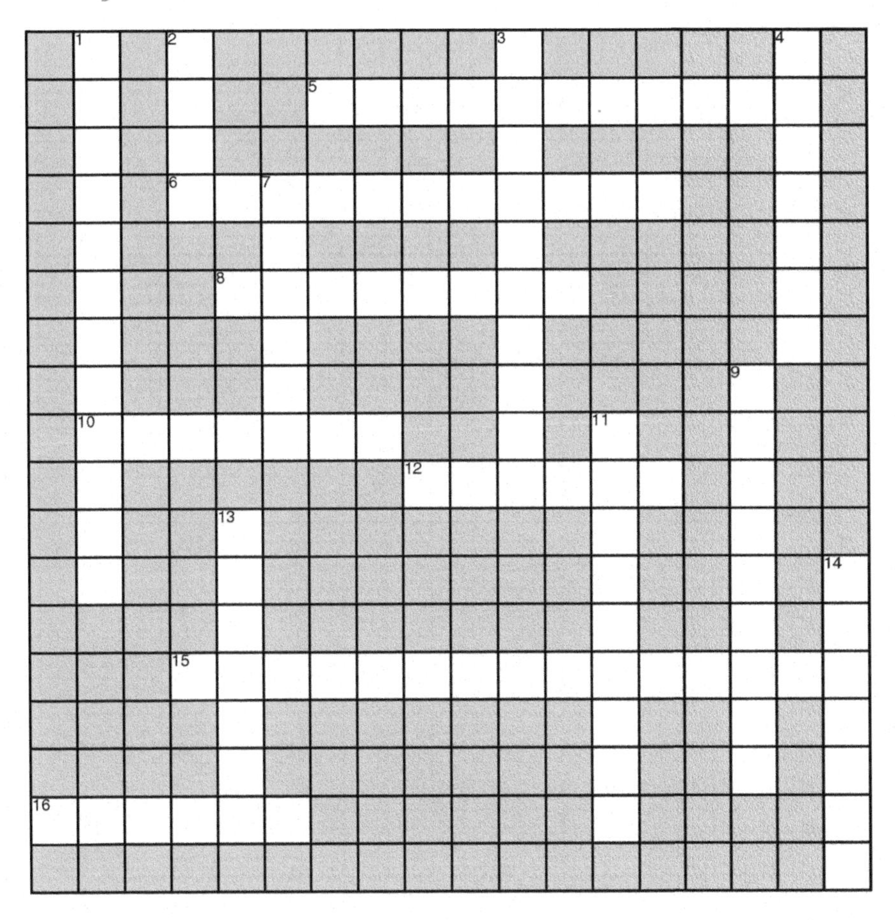

Clues Across

5. Stress hormone.
6. The division of the Autonomic Nervous System that prepares the body for fight or flight.
8. Membranes which protect the Central Nervous System.
10. Glands where stress hormones are produced.
12. System in charge of emotions.
15. This system restores the body to a resting state.
16. Outer layer of the brain.

Clues Down

1. This is the size of a pea and controls body temperature.
2. Joins the hemispheres of the cerebellum and connects the cerebrum to the cerebellum.
3. Regulates balance, movement and muscle coordination.
4. Another word for nerve cell.
7. Sheath covering the axon.
9. Nerve fibres which transmit impulses to the cell body.
11. Gland at base of brain which secretes hormones.
13. Where one neurone meets another.
14. Centre of a nerve cell.

The reproductive system crossword

(answers on page 175)

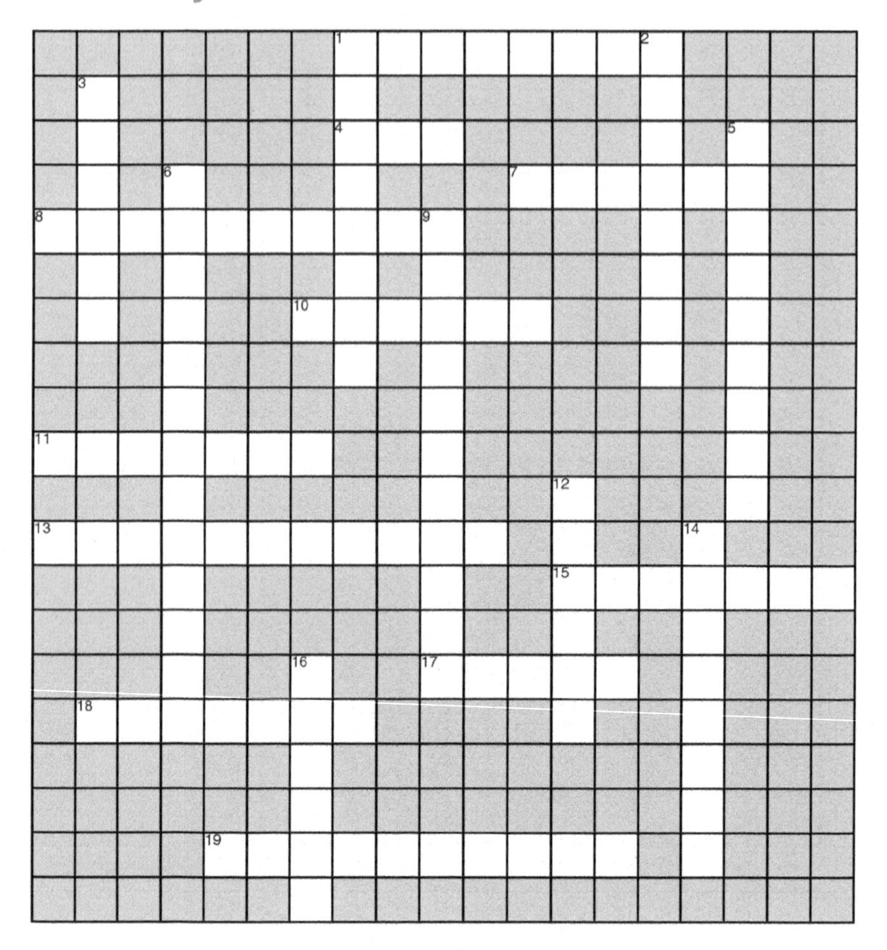

Clues Across

1. The unborn baby's support system.
4. The eggs that the female reproductive system produces.
7. This muscular passage connects the cervix to the vulva.
8. A type of ovarian syndrome.
10. The fertilized ovum grows into a baby here.
11. A pregnancy that occurs outside the uterus.
13. These are produced and stored in the testes.
15. The sac of skin and muscle where the testes are contained.
17. The fluid ejaculated during intercourse.
18. Glands that secrete oestrogen and progesterone.
19. A tightly coiled tube which transports sperm.

Clues Down

1. A small gland between the bladder and rectum in men.
2. This fluid protects the unborn baby from shocks.
3. The first cell of the baby.
5. The uterus connects to these tubes.
6. Very painful menstruation.
9. The nucleus in the head of a sperm contains 23 of these.
12. Glands contained within the scrotum.
14. This piece of skin is sometimes removed from the penis for religious or hygienic reasons.
16. The neck of the womb.

The skeletal system crossword

(answers on page 176)

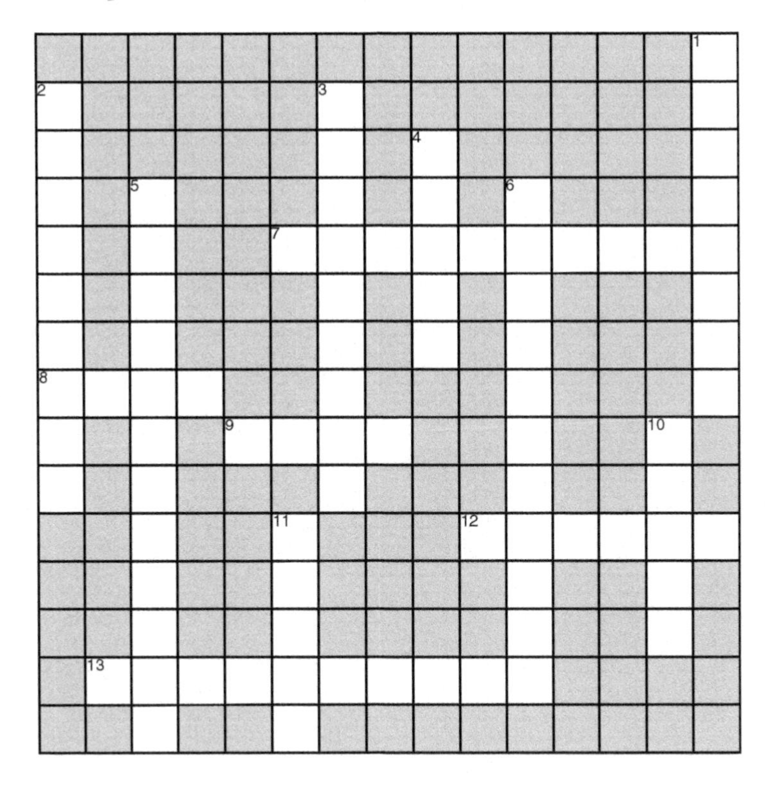

Clues Across

7. White skin-like covering of most bones.
8. Type of bone that allows the movement of the limbs.
9. Bones with broad surfaces for muscle attachment, such as the shoulder blades.
12. This is where fats are stored in the bone.
13. This type of tissue looks like a sponge.

Clues Down

1. The type of bone that makes up the knee-cap.
2. All the bones of the face are like this.
3. These canals run through the bone tissue.
4. This skeleton supports the head, neck and torso.
5. This type of skeleton supports the limbs.
6. Bones are made of these cells.
10. Strong bones but where little movement is required.
11. 206 of these make up the skeleton.

The skin crossword

(answers on page 176)

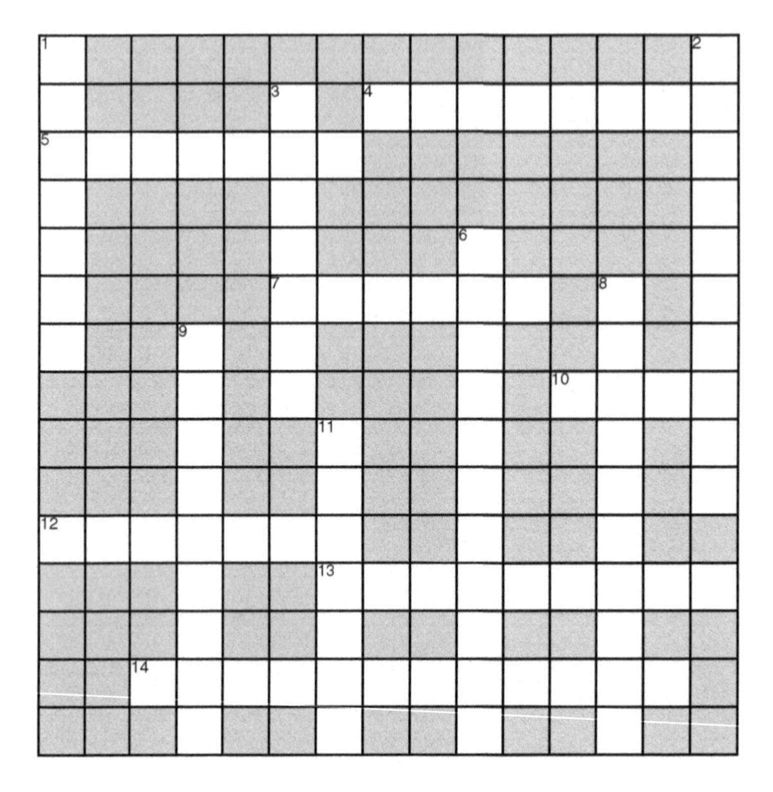

Clues Across

4. These sweat glands produce a fluid which creates body odour when mixed with bacteria on the skin surface.
5. The top surface of the epidermis.
7. Lies beneath epidermis and contains sweat glands.
10. These cells produce histamine in response to an allergic reaction.
12. These conical projections contain blood vessels and nerves which supply the hair with nutrients.
13. These glands produce sebum.
14. Bottom layer of epidermis, containing melanin pigment of the skin.

Clues Down

1. These sweat glands help control body temperature.
2. These are white blood cells that fight disease and infection.
3. The second clear layer of the epidermis.
6. These are cells responsible for the production of collagen, elastin and areolar tissue.
8. The third, granular layer of the epidermis.
9. These contain hair.
11. The dermis contains collagen and this kind of tissue which keeps the skin supple and elastic.

The urinary system crossword

(answers on page 177)

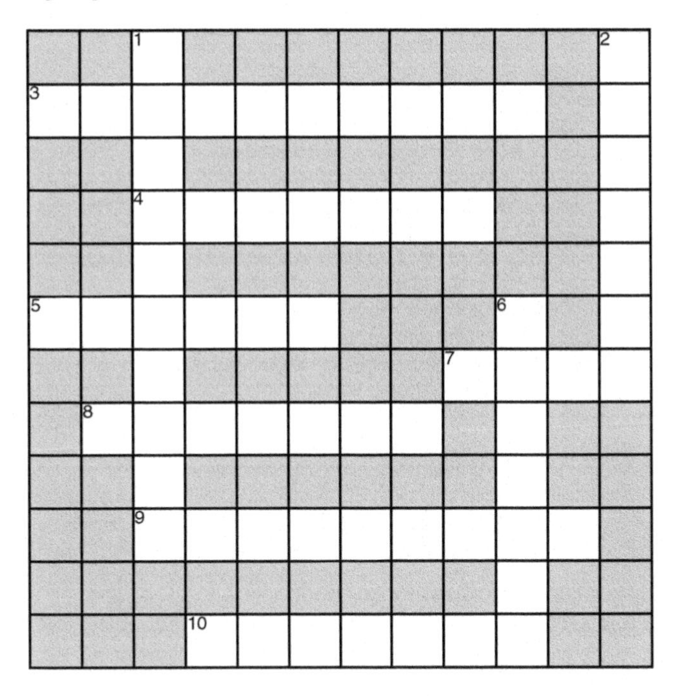

Clues Across

3. This takes place in the Bowman's capsule.
4. This is the inside part of the kidney.
5. This is the outside of the kidney.
7. Urine contains about 2% of this substance.
8. This holds urine.
9. This relaxes when the bladder wall contracts, emptying the urine into the urethra.
10. Bean-shaped organs that filter toxic substances from blood.

Clues Down

1. Capillaries surrounded by the Bowman's capsule.
2. Tube from bladder to outside of body. It is the passage for semen in men.
6. Tubes connecting kidneys to bladder.

Chapter 9

Read right

OVERVIEW

This chapter will help you:

- ✓ **understand how to improve your reading of challenging A&P texts;**
- ✓ **make reading easier for yourself;**
- ✓ **discover the 'right' speed for reading;**
- ✓ **make the best possible notes.**

Textbooks and supplementary reading

Make friends with your textbooks, not enemies . . . it is easy to become anxious about the amount of technical reading required when studying Anatomy and Physiology. At this level of study, you will be given reading tasks and expected to become an independent learner.

Your textbooks and supplementary handouts should be essential tools (not obstacles) to help you succeed in your studies. If you read material in preparation for a class, you can then use contact time to understand complex processes, rather than trying to grasp or record the basic facts. On the other hand, 'reading round' a subject after a lecture, at your own pace, gives you more control: it should help you understand the topic better.

It is very common, however, to have difficulties fitting reading around lectures, writing, practical work, revising notes, and everyday life! There is no one right way to read: just ways that are right for you and the particular task you have in hand. In this section, you will find ways to make reading complex texts easier.

Why can reading about Anatomy and Physiology be challenging?

1. At first, it can be hard to deal with the
   - **new technical vocabulary**;
   - **detailed abstract concepts**;
   - and **multi-step processes**.

   This can be even more daunting, if you have limited scientific knowledge.

2. **Some texts are easier to read than others**. Let's face it: not all A&P texts are clear, simple, well organized, well presented and well illustrated.

3. **You may have problems with your eyesight**: reading may be tiring and physically uncomfortable for you. You may find that print blurs or moves. You may find that you frequently miss a line, re-read the same bit or misread some text. This may happen more often when print is small, cramped or against a bright white background. A small number of people may have a Specific Learning Difficulty, such as dyslexia. This can affect their accuracy, speed or memory.

4. Your **concentration may be affected** when you are tired, stressed or unwell. A distracting, noisy or uncomfortable environment can affect your ability to absorb text.

5. You may **lack an effective reading and note-making strategy**.

There are ways to remedy all these challenges: they are explored below.

What is the 'right' speed to read?

There is no 'right' speed that you should be reading at. The average reading speed is about 200 to 250 words per minute. That means that reading 100 words should take you between 30 and 45 seconds! But skimming through a simple 100-word newspaper article will take you less time than studying a complex academic text of the same length. Try this exercise:

How to work out your own reading speed

- Prepare to answer these questions *after* reading 'Text A' on page 136.
  - At what time does Tom Green usually take his dog for a walk?
  - What did he see on arrival at the beach?
  - What happened to the young couple?
- Prepare to time yourself reading the text.
- Read the text *silently*.
- Write down the time you took.

TEXT A Suicide couple found on beach

Last Friday night, Tom Green, from East Sands, came across a sight he will never forget. As usual, he took his dog for a last walk at around 11:30 that night. But on that occasion, for some reason, he decided to head for a lonely beach he rarely visited. On arrival, he saw a man and a woman on the sand. When he got closer, he saw that they looked about 15 or 16 years old. They walked hand in hand in the moonlight. Then they walked right into the water. The young couple were rescued and taken to hospital.

(100 words)

- Prepare to answer these questions *after* reading the text:
  - What is near the centre of the cell body?
  - What do neurofibrils extend into?
  - What is the structure of a neuron?
- Prepare to time yourself reading the text.
- Read the text *silently*.
- Write down the time you took.

TEXT B Neuron structure

Neurons vary considerably both in size and shape; but they all have common features. The neuron cell body consists of granular cytoplasm, a cell membrane, and organelles such as mitochondria, lysosomes, a Golgi apparatus, and a network of fine threads called neurofibrils, which extend into the axons. Scattered throughout the cytoplasm are many membranous sacs called chromatophilic substance (Nissl bodies), which are similar to rough endoplasmic reticulum in other cells. Ribosomes attached to chromatophilic substance function in protein synthesis, as they do elsewhere. Near the centre of the cell body there is a large, spherical nucleus with a conspicuous nucleolus.

(100 words, taken from *Hole's Essentials of Human Anatomy & Physiology*, published by McGraw-Hill)

How long did it take you to read 'Text A' in contrast to 'Text B'?

What was your average speed?

Do you think you would need to re-read 'Text B' several times to answer the questions correctly?

If you want to increase your speed, and make reading easier, you will find some solutions in the following pages.

How can you make reading easier (or harder)?

There are 10 main factors that will affect your reading speed:

1. the type of text that you are reading (academic, legal document, magazine, dictionary etc.);
2. your reason for reading (study, pleasure, reference etc.);
3. lack of familiarity with the subject matter that you are reading about;
4. sight difficulties;
5. inability to concentrate due to tiredness or stress; or being interrupted by noise or other people;
6. length of sentences (over 20 words becomes challenging);
7. text containing long, unfamiliar, technical words, and explanations that are unclear;
8. text that is poorly organized;
9. text that is printed in small cramped print or with few helpful illustrations;
10. text printed on very white shiny paper.

Reading technical Anatomy & Physiology books is challenging. But you can have control over most of the factors listed above. Below, you will find five main suggestions for making reading easier and more effective for you.

Choose the most 'readable' textbook

Readability is the sum total of all those elements within a piece of writing that affect the success the reader can have with it. In the section above, you learnt about the 10 main factors that will affect your reading speed. Five of these factors relate to the *style of the text* you are reading, such as:

- the length of sentences;
- long, unfamiliar, technical words;
- confusing explanations;
- how the text is organized;
- what the print looks like on the page.

Usually your tutors will have recommended particular textbooks, articles or handouts for you. If you really cannot get on with the style of a text, find another way of learning the topic.

- Ask your librarian, tutor or fellow students to recommend a more readable text.
- Find an alternative source of information: perhaps a CD ROM or internet site.
- Scan the text into a computer and use software to change the appearance of the text or have it read out aloud to you. See page 55 for IT solutions that can help you with your reading.
- You can also use a PC to work out the 'readability' of a particular text. Microsoft® Word uses the *Flesch Reading Ease readability score formula.* This rates a text on a

100-point scale based on the average number of syllables per word and words per sentence. The higher the Flesch Reading Ease score, the easier it might be to understand the document. Aim for a Flesch Reading Ease score of approximately 60 to 70. For example, Text A, above, has a score of 74.8. Text B has a score of 31.1. No wonder Text B takes much longer to read!

To get a readability score, use your 'Review' menu. Select 'Spelling and Grammar', then 'Options', then 'Show Readability Statistics'. Then each time you check the spelling of a text, the Readability score will appear along with a word count.

2 *Make sure you are in the right physical and mental state*

Reading about detailed and complicated scientific ideas requires concentration.

- Be aware that tiredness, stress and distractions will affect your reading performance.
- It's better to read efficiently in small bursts, than just 'get it done' for the sake of it and retain nothing.
- If necessary, get your eyesight tested by an optometrist. Ask to be tested for 'colour sensitivity' if:
  - the whiteness of the page seems to dazzle you;
  - your eyes get very tired;
  - you get frequent headaches;
  - the print begins to move or blur;
  - you keep losing your place.

If you suspect you might have an unusual or persistent reading difficulty, speak to your tutor or contact the British Dyslexia Association for advice.

3 *Vary your reading style to suit your purpose*

- If you just want to get a general idea, don't read every word. Lightly **skim** your eyes over the text.
- If you just want to find a particular piece of information, don't read every word. Just train your eyes to **scan** quickly over text.
- If you need to really understand a text, you will need to **read slowly** and **actively search for meaning**.
- If you want to memorize a text, you will need to read with a pencil in hand, **underlining key points and making notes**.
- If you are doing research, you need to **critically evaluate** the argument presented by the writer. You may also need to question the value of their evidence.

Why bother using a Reading strategy?

Do you ever read a chapter in a textbook and then find that you can't remember much about what you have just read? If you just get your reading over and done with, without learning anything, you are wasting a lot of your precious time and energy. There is a way to remedy this: you can train your mind to learn as you read. Just use the '**SWIMMER**' strategy, which is explained below. It may take you longer than simply reading a text, but in the long run it will save you time and prevent stress.

What are the advantages of using the 'SWIMMER' strategy?

Students who already use the SWIMMER strategy report these definite advantages:

- You can read a bit at a time. That way you make good use of short periods of time you might normally waste and you don't overload your brain.
- Using a strategy helps you stay focused and avoid boredom.
- You shouldn't need to re-read the chapter you have worked on, thus saving time in the long run.
- You can't fool yourself about what you know or don't know about what you have just read. It should be a thorough way of testing that you really understand the material you are studying: this will make it much easier to fix it in your long-term memory and recall it later.
- You will have ready-made notes, when it comes to exam revision or writing an essay.
- You will have practised test questions in advance.

What does the SWIMMER strategy mean?

Each letter in '**SWIMMER**' helps you remember a stage in the overall strategy sequence. **S-W-I-M-M-E-R** stands for:

Skim-scan;
Write questions down;
Intensive reading;
Make notes;
Memory check;
Evaluate;
Revise.

The strategy is broken down into 7 steps and these are explained on pages 140–1.

To help yourself remember the word SWIMMER, imagine this: you could find yourself in deep water and drowning in all the reading you have to do. So you need to become a strong swimmer, to stay afloat and move smoothly towards your goals.

The SWIMMER strategy explained

1 **Skim-scan the chapter quickly** – glance over but do not read every word of the whole chapter: this will warm up your brain in preparation for a more intensive read later on.

- Read the title and any introduction – this will help you get an idea of the general content and main points of what you are about to learn.
- Read any heading and subheadings in large or bold print– this will give you an idea of how the information is organized.
- Read the topic sentence – usually the first one in the paragraph. This will introduce the main point.
- Notice any graphics – pictures, charts, maps, diagrams etc. – these should help make the information easier to understand and more memorable.
- Read any summary or review questions – this will give you an idea of what the author thinks are the most important points.

2 **Write questions down** – this will keep your mind actively searching for answers as you read later on.

- Take your first paragraph, or section under a heading, and write down any questions you hope might be answered in that section. It helps to put yourself in the shoes of a lecturer and predict what possible questions he/she might ask about the information contained in the first section.
- You can write in pencil in the margin of your text or on a separate piece of paper.
- Don't worry if you can't find more than one or two questions to ask: you can always add more questions at a later stage, if you need to, when you have read the paragraph.

3 **Intensive reading** – reading for meaning.

- Slowly and carefully read the section with your questions in mind. Look for the answers, and notice if you need to make up some new questions.
- If you find your mind wandering at any point, read aloud or take a rest.
- *Don't make notes of key points during this first reading.* Why is this better for learning? If you highlight or take notes during your first reading, most of what you read seems to contain important information. You don't have the overall picture of that section, so it's easy to confuse less important details with the important points. You will end up with too much of the section highlighted or too many notes. Remember, the purpose of underlining or highlighting your text, or making notes, is to summarize the material for later revision. When you revise for an exam or an essay, you do not want to read the text again in order to select the important information.

4 **Make notes** of key points, important terminology and useful explanations.

- Invent your own system for highlighting, underlining, numbering, circling, summarizing in your own words or illustrating information.
- Have a medical dictionary on your desk to look up any new technical words. Skipping unknown words in the hope of 'guessing' them through context reduces your understanding of scientific terms. Write any new word down in a personal notebook that you can refer to later if you need it.

5 **Memory check**

- Cover up the text and ask yourself the questions you wrote down before.
- Check if you can answer them in your own words.
- Reciting your answers aloud or explaining the concepts to another person will fix the information very firmly in your memory.

6 **Evaluate**

- How did you do? If you can't answer the questions you set yourself, look back over the text again – as often as necessary.
- You may also want to add some more or different questions at this stage.
- Don't go on to the next section until you feel that you can answer the questions confidently.

7 **Review**

- When you've finished the whole chapter using the preceding steps, go back over all the questions. See if you can still answer them. If not, look back and refresh your memory.
- Do any practice questions and exercises that the author or your lecturer has set.
- Check your notes:

  1 Are they clear and helpful for future revision?
  2 Do they need recopying or redoing?
  3 Would a diagram or Mind Map help you remember them better? See note-taking suggestions below.

If you don't actively revise, you will probably forget 80 per cent of what you read within just two weeks. If you want to move information from your short-term into your long-term memory, you will need to review information several times in short bursts. There are many suggested memory strategies in this book to help you.

5 Make the best notes possible

What skills are required for note-taking in Physiology?

Many concepts in Physiology revolve around multi-step sequences or a cycle of events. This lends itself to a specific type of reading and note-taking approach, as outlined later.

- Look out for 'signpost words' which signal the beginning, middle, end or main parts of a sequence.
- Look for 'signpost words' which signal relationships between ideas. Words like 'when', 'if', 'since', 'although', 'as', 'because', 'in turn', 'as a result', 'consequently', 'finally' etc.
- Break the reading of long sequences into a smaller numbers of steps.
- Sequence and number the key steps in a process or cycle.
- Create flowcharts, diagrams and sketches to show a visual representation of written steps: this can increase memory of sequences.

How can you save time *and* make good notes?

Good note-making saves you a lot of time: you only focus on the *bare minimum needed.* This gives you a fast way to revise for exams or essays. It also helps information stick.

1. Note only main ideas, topic sentences, and key words.
   Highlighting nearly every sentence or word on a page clearly means that you don't understand the text.
2. Make summaries of only key points ('skeleton notes') in your own words.
   You may wish to write in pencil in the margin of the page.
   Later you can recopy your notes on paper, index cards or a PC.
3. Draw diagrams of your notes.
   This helps understanding and fixes information in your memory.
4. Visually organize points into groups.
   Use headings, sub-headings, numbering, lists with bullet points, boxes, etc.
   This method forces you to pay attention to the structure of the material.
5. Use coloured pens or highlighters to mark different concepts – such as key terms, definitions, examples, etc.
6. Use your own abbreviations and symbols – such as arrows, stars, questions marks.
7. Create a personal mark-up system – find what works for *you* and stick to it!
   For example:
   - highlight in colour a very important detail, such as a topic sentence (a 'topic sentence' introduces the main idea of a paragraph);
   - underline key points or words;
   - put a question mark by new words or points you don't understand;
   - put a star next to things you find unusual or interesting.

Activity: Making notes

- Mark up the text below.
- Next make skeleton notes of it.
- Compare what you have done with the examples on page 144.

THE ENDOCRINE SYSTEM

The Endocrine System consists of ductless glands that produce hormones. Hormones are chemical messengers, which control the activities of all the body processes. When hormones are secreted, they go straight into the bloodstream, from where they are transported to all parts of the body. Each gland produces specific hormones. For example, the pancreas secretes insulin in the islets of Langerhans. Insulin helps glucose enter cells, thereby regulating blood sugar levels. After eating, the level of glucose increases in the blood. This increase is detected by receptors, which tell the pancreas to release insulin. This hormone travels in the bloodstream to the liver, where it causes the liver to take up glucose from the blood: this then brings the glucose level in the blood back to normal. The liver stores the extra glucose as glycogen. If a person does not secrete insulin, the level of glucose does not decrease in the blood, leading to a condition known as diabetes mellitus.

Example of text mark-up

THE ENDOCRINE SYSTEM

The Endocrine System consists of ductless glands that produce hormones. Hormones are chemical messengers, which control the activities of all the body processes. When hormones are secreted, they go straight into the bloodstream, from where they are transported to all parts of the body. Each gland produces specific hormones. For example, the pancreas secretes insulin in the islets of ? Langerhans. Insulin helps glucose enter cells, thereby regulating blood sugar levels. After eating, the level of glucose increases in the blood. This increase is detected by receptors, which tell the pancreas to release insulin. This hormone travels in the bloodstream to the liver, where it causes the liver to take up glucose from the blood: this then brings the glucose level in the blood back to normal. The liver stores the extra glucose as glycogen. If a person does not secrete insulin, the level of glucose does not decrease in the blood, leading to a condition known as diabetes mellitus.*

Example of skeleton notes

ENDOCRINE SYSTEM

A. Composition: ductless glands producing hormones
B. Function:

1 **Hormones:**
- chemical messengers
- control all body processes
- go into bloodstream ⟶ all body parts.

2 **Glands**: each produces specific hormones.
E.g. Pancreas:
- Secretes insulin in islets of Langerhans.
- Insulin helps glucose enter cells ⟶ regulated blood sugar levels.
- How?
  . After eating: glucose level increases in blood ⟶
  . receptors detect increase ⟶
  . pancreas releases insulin ⟶ in bloodstream to liver ⟶
  . liver to take up glucose from the blood.
  . The glucose level in blood ⟶ normal
  . liver stores extra glucose as glycogen.

* *If person doesn't secrete insulin: level of glucose doesn't decrease in blood ⟶ diabetes mellitus.*

Chapter 10

Cool calculations

OVERVIEW

In this chapter, you will have the opportunity to explore:

- ✓ **why people are afraid of figures;**
- ✓ **why bother improving your maths skills;**
- ✓ **how to make an Action Plan that you stick to;**
- ✓ **how to identify precisely what maths you will need to know;**
- ✓ **how to test yourself to find out your strengths and weakness in maths;**
- ✓ **how to work through some maths problems, including drug calculations;**
- ✓ **some handy tips for calculations, including decimals and conversions;**
- ✓ **the value of keeping a Maths Emergency Notebook;**
- ✓ **some resources that other students have found useful.**

Are you frightened of figures?

How does maths sneak into nearly everything we do?

All day long we solve mathematical problems without it being a big deal. We roughly estimate whether we can afford a holiday; how much coffee we have left in a jar; whether we have time to have a shower before a favourite programme starts; what the sex of a baby might be; how likely it is that a couple might split up, etc. We also have to make precise calculations such as how much change to give somebody; what time to leave home to catch a train; how long to leave a cup of soup in the microwave before it explodes! Maths can also help us work out how to meet the perfect partner; how politicians can 'lie' to us by manipulating figures; how likely it is that you will catch 'flu; or why weather forecasters often get it wrong . . . Give us a problem we care enough about and we can all be keen mathematicians!

So, why are so many people frightened of figures?

Many people found school maths scary, confusing or mind-numbingly boring. As adults, we find these feelings often come back when we are in a situation that requires calculations. Under stress, your maths skills can appear to break down completely: for example, when you have to do a calculation in front of others, when you need to solve a problem quickly, when you are very tired or if the environment is noisy. The pressure is even greater if you think that a small error in a calculation could have a serious effect on a patient/client's wellbeing or on your own career.

Use it or lose it . . .

Many people use computers and calculators to avoid difficulties with calculations. However, often errors happen because the wrong data has been entered and not detected. You can prevent errors of this type by having another member of staff check your results, or you can do a pen-and-paper calculation to double check. As the use of computers and calculators can cause you to get rusty, it is therefore important that you can estimate answers and perform the calculations on paper so as to maintain your maths skills.

Does any of this ring true for you?

So if you wish to improve your maths skills for work, you will find in this chapter suggestions that other students have found very useful.

The first step is to work out what maths you will actually need at work: see page 147.

Then, you can make an Action Plan to address any difficulties you have with making calculations: see page 148.

You are far more likely to carry out your resolutions if you:

- **write them down;**
- **have clear goals;**
- **break these goals into small steps;**
- **and tell others about your resolutions so that they can keep you on track.**

What Maths will *you* need at work?

Use the diagram below to decide what areas you need to focus on.

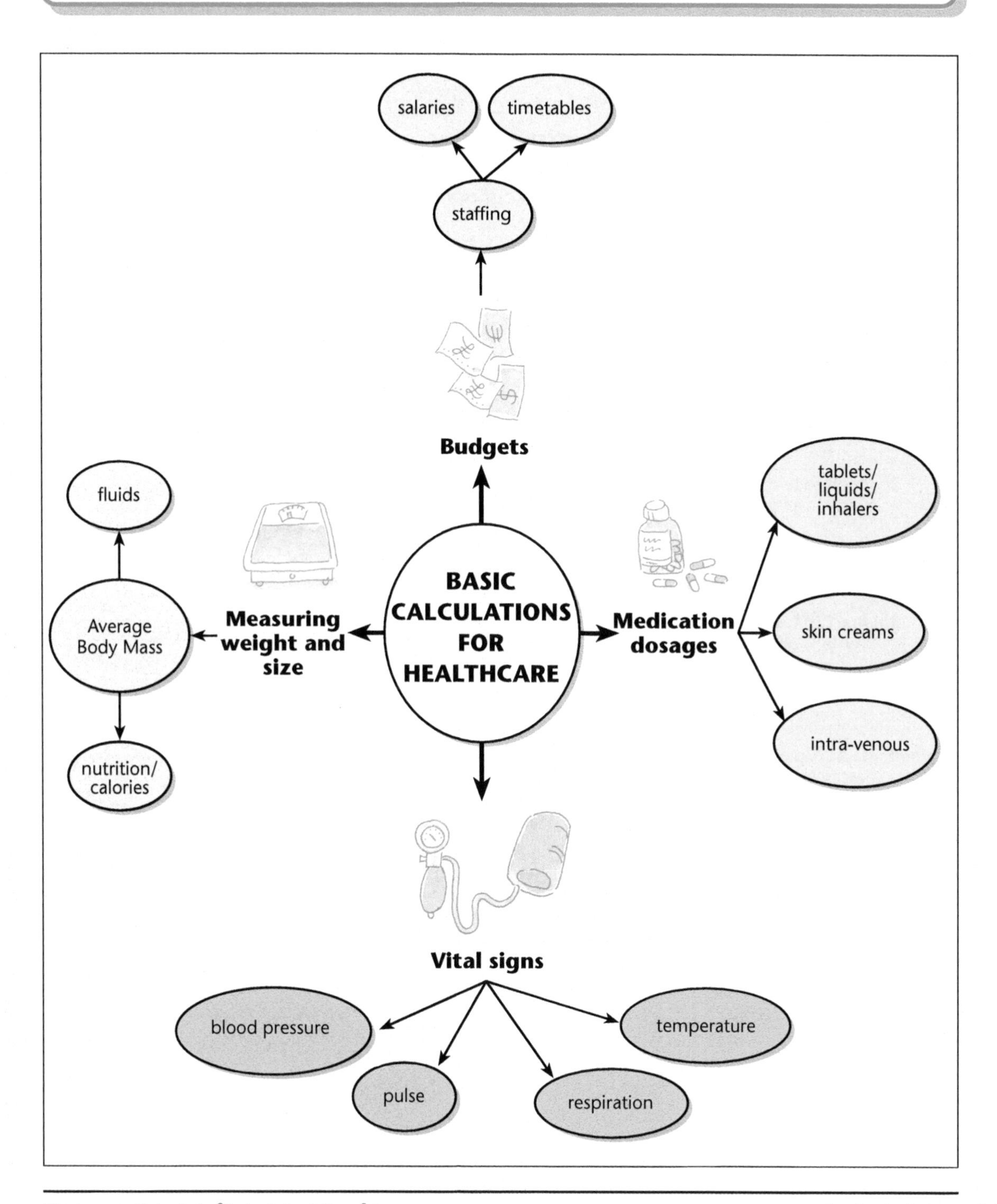

Diagram created in Inspiration® by Inspiration Software®, Inc.

Activity: Maths Action Plan

Read one student's Action Plan below. She is determined to tackle her maths problems. Then write your own Action Plan and show it to somebody who will keep you on track!

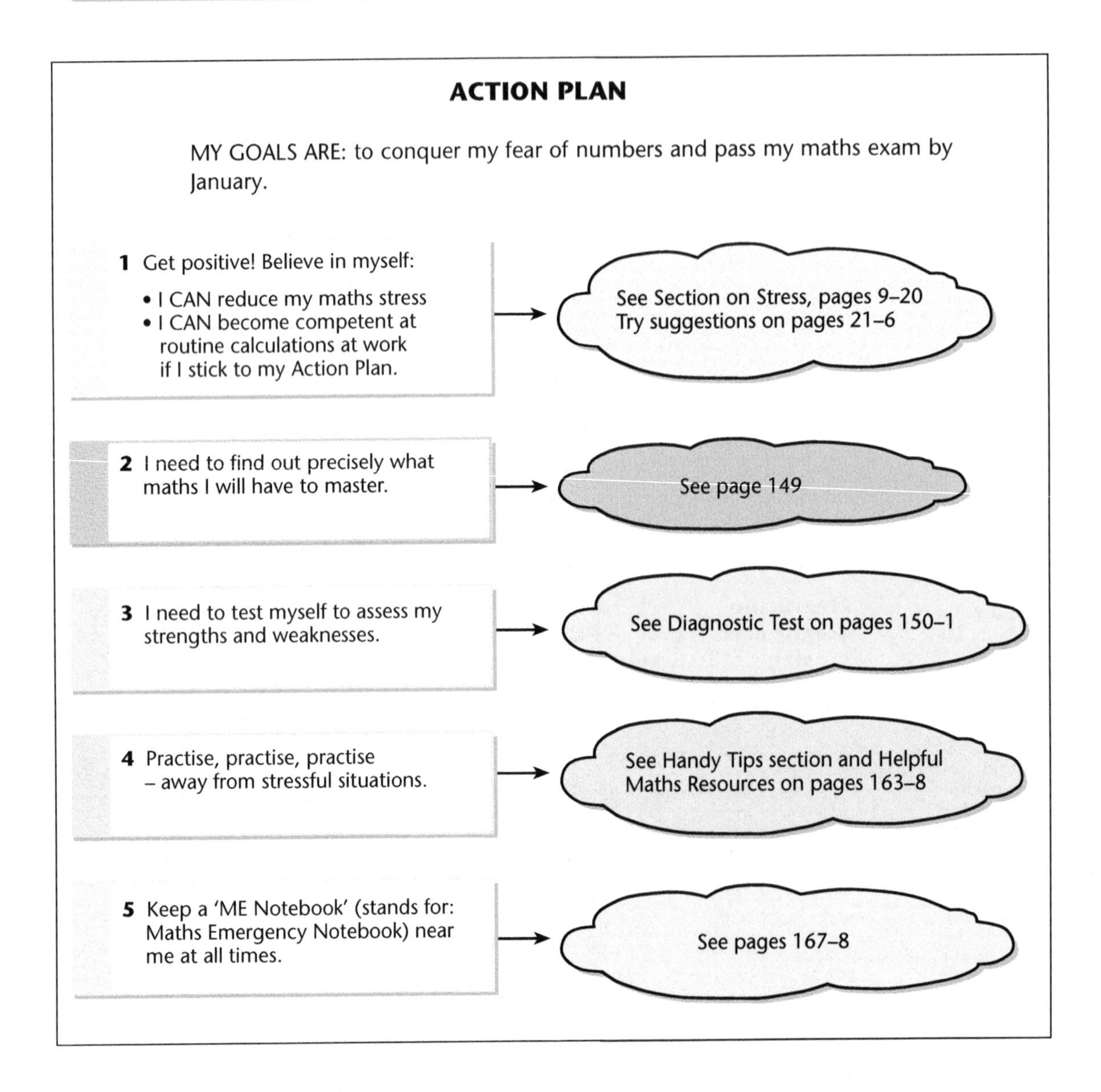

Diagram created in Inspiration® by Inspiration Software®, Inc.

What maths skills do you need to revise?

Use the Test on pages 150–1 and work through the answers.
Then, using the diagram below, identify what maths skills you need to revise.

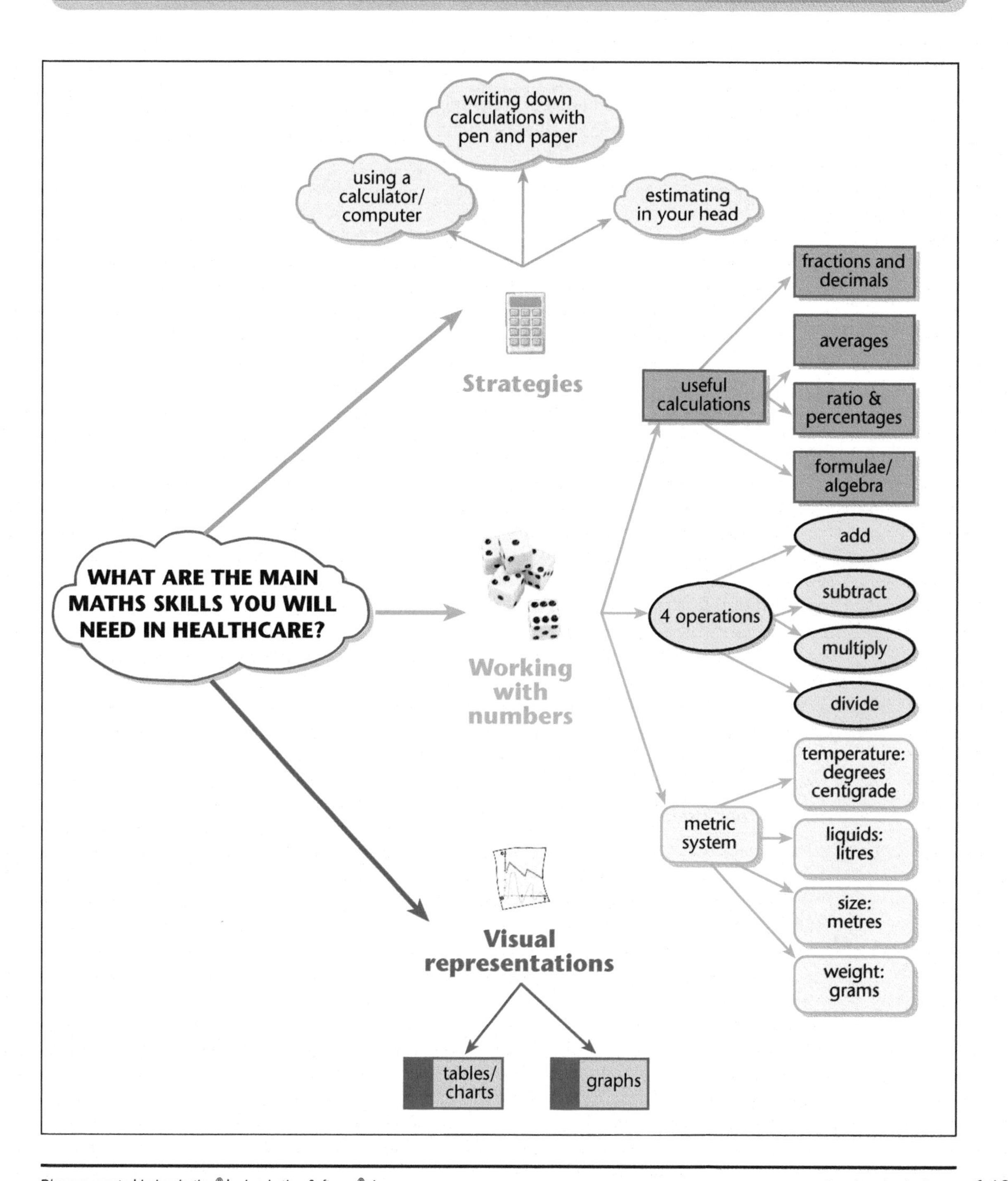

Diagram created in Inspiration® by Inspiration Software®, Inc.

Test Yourself: How cool are your calculations?

Question 1

Jim has a temperature of 38.8°C. His normal temperature should be about 37°C in the mouth. How much is his temperature over the norm? Should you be concerned?

Question 2

You have a bottle of 50 tablets each containing 300 mg of aspirin. You need a dose of 450 mg to be taken 3 times a day. How many tablets in total do you need for 4 days?

Question 3

Sean usually mixes his own massage oils for aromatherapy patients. The usual proportion is 8 drops to 20 ml of almond oil. How many drops of lavender does Sean need to add for 5 ml of almond oil?

Question 4

You can buy ten tubes of eye ointment of 2.7 g each or one hundred tubes of 26.8 mg each. Can you estimate which quantity is larger? Check your estimate with pen and paper and then with a calculator.

Question 5

An osteopath's practice opens at 9 a.m. Ann's appointment is at 9 a.m. for 30 minutes. John's is at 9.30 for 20 minutes. Moira's at 9.50 for 5 minutes. Ali's at 9.55 for 10 minutes. Moira and Ali have to get to work as early as possible. How could the appointments be re-scheduled so that waiting times are reduced?

Question 6

Kelly is allowed a maximum of 1 litre of fluids per day. She has had: 3 cups of tea of 155 ml each; 1 glass of orange juice of 178 ml; 108 ml of water. Can she have another glass of juice? Estimate the answer in your head without working it out exactly. Then use a calculator to check you are right.

Question 7

Your salary is £24,000. You have been offered a pay rise of 2.5%. How much will your salary be?

Question 8

Sally's weight is 77.5 kg and she is 1.55 m tall. What is her Body Mass Index?

$$\text{BMI} = \frac{\text{weight (kg)}}{\text{height (m) squared}}$$

Question 9

Look at the table relating to BMI, below. Is Sally's weight a cause for concern?

Below 20	underweight
20–24.9	ideal
25–29.9	overweight
30–39.9	obese
Over 40	extremely obese

Question 10

Look at Emma's pulse rate chart. 60–80 beats per minute is average for adults. How long does it take for her pulse to return to its normal resting rate after she stops exercising?

Pulse rate after exercise

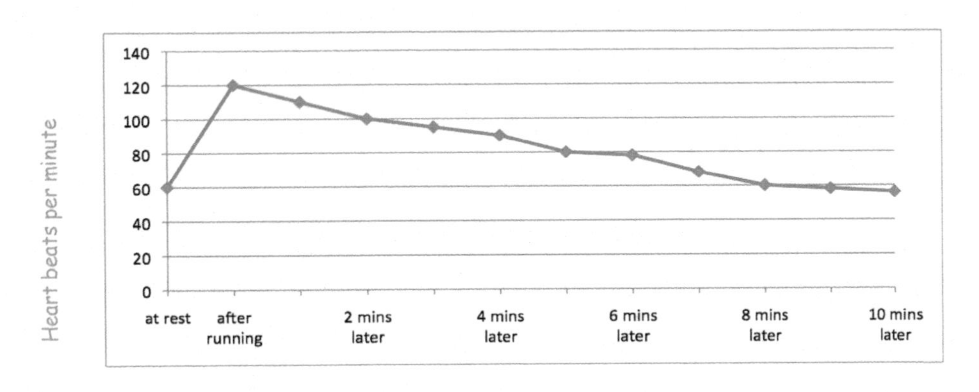

Elapsed time

(Answers on pages 153–62)

If you are not sure how to start tackling a question, use the Helpsheet on page 152.

Helpsheet: Working through the steps of a mathematical problem

Make a copy of the form on the next page to help you work through the steps of a mathematical problem.

Write your answers in the blank boxes.

Helpsheet

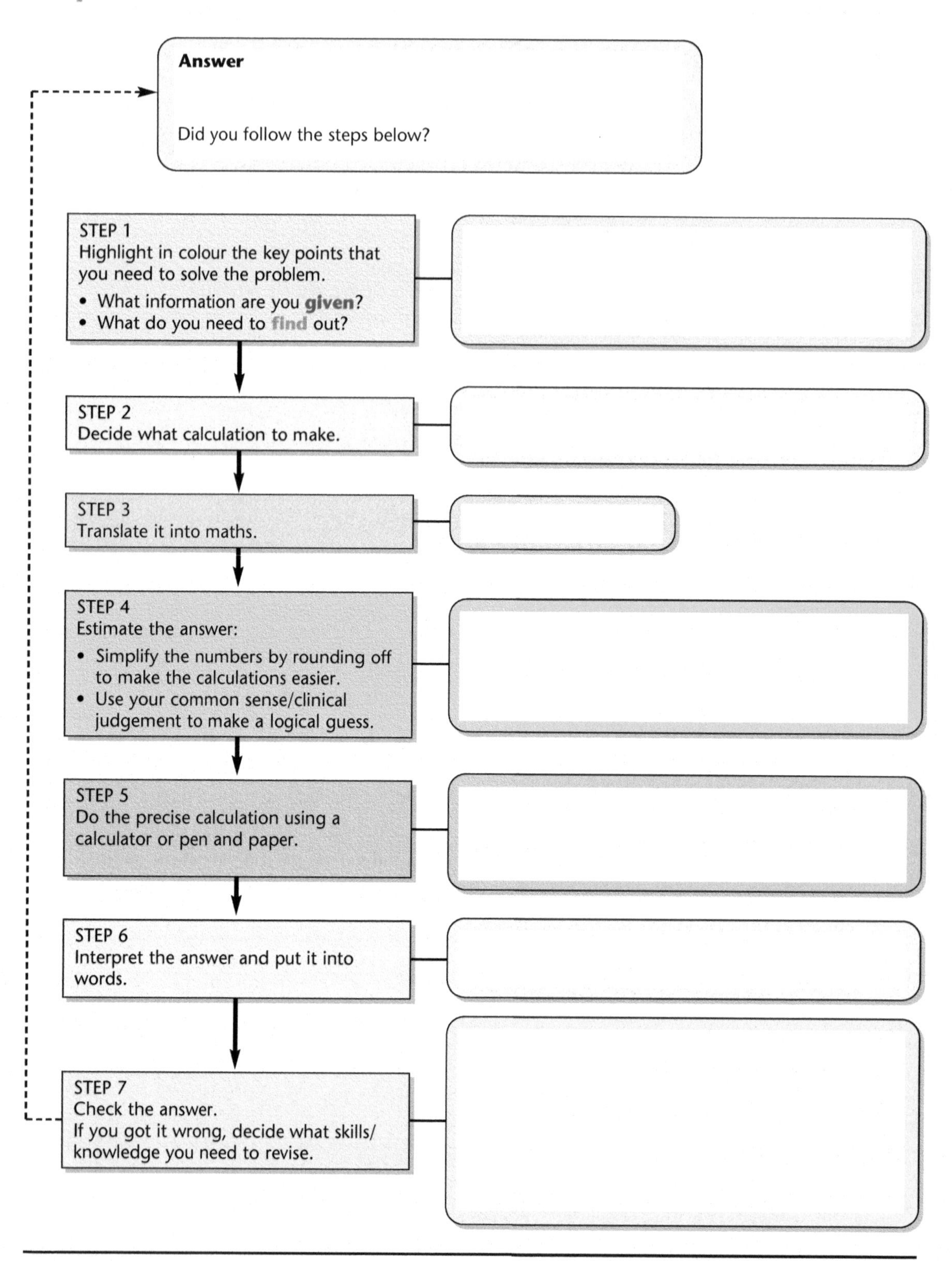

Diagram created in Inspiration® by Inspiration Software®, Inc.

Question 1

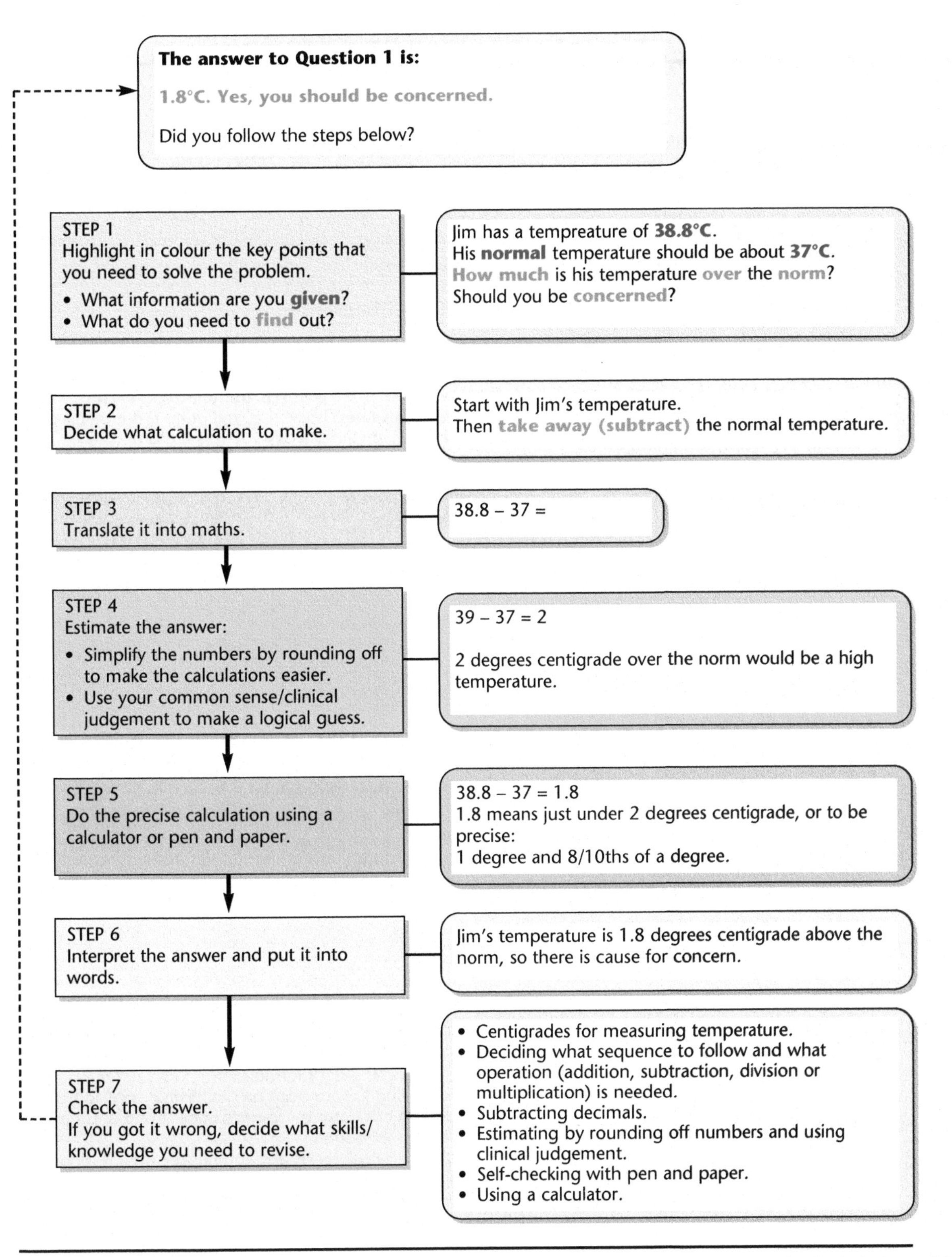

Diagram created in Inspiration® by Inspiration Software®, Inc.

Question 2

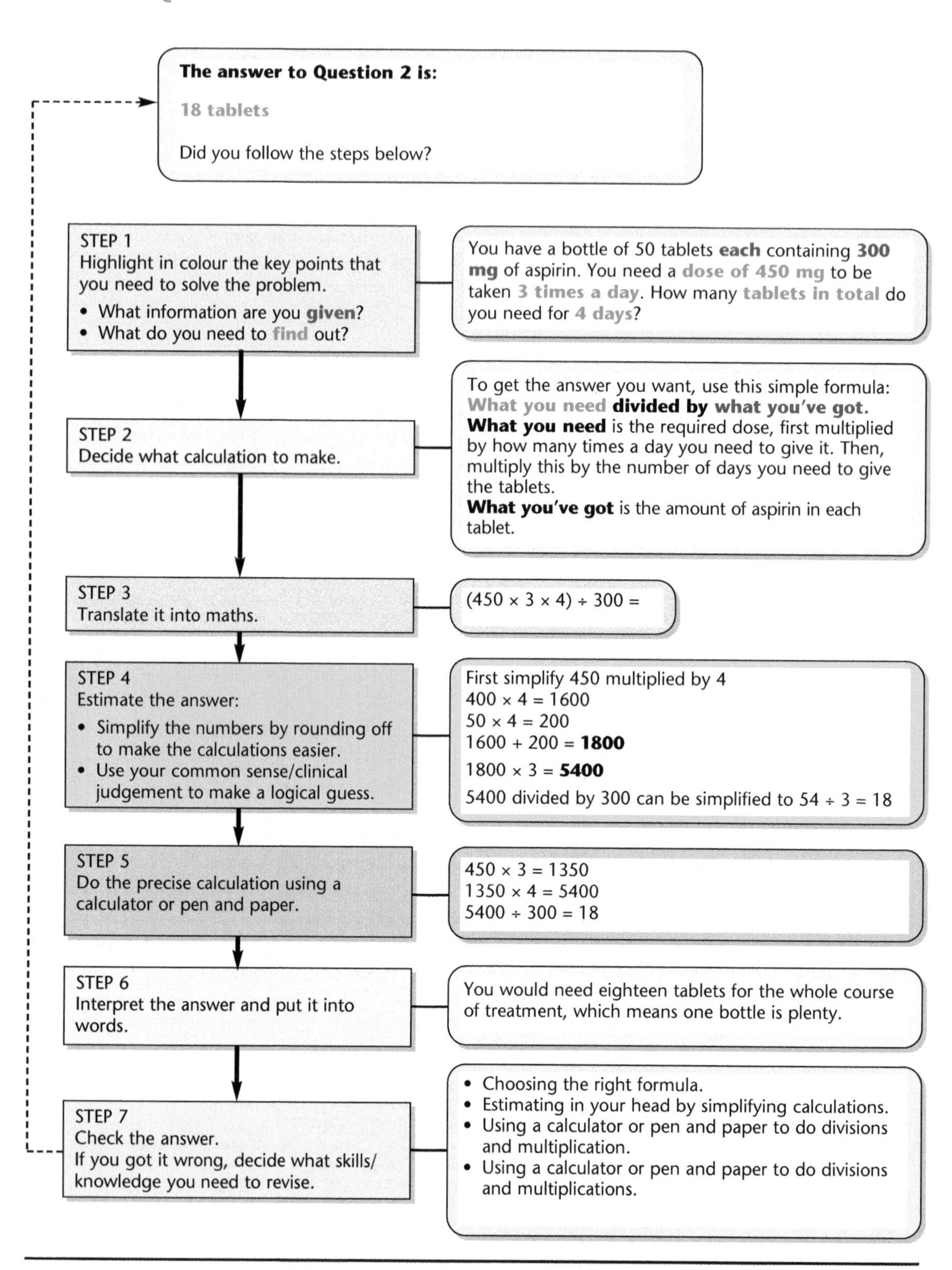
The answer to Question 2 is:
18 tablets
Did you follow the steps below?
STEP 1
Highlight in colour the key points that you need to solve the problem.
• What information are you given?
• What do you need to find out?
You have a bottle of 50 tablets each containing 300 mg of aspirin. You need a dose of 450 mg to be taken 3 times a day. How many tablets in total do you need for 4 days?
STEP 2
Decide what calculation to make.
To get the answer you want, use this simple formula:
What you need divided by what you've got.
What you need is the required dose, first multiplied by how many times a day you need to give it. Then, multiply this by the number of days you need to give the tablets.
What you've got is the amount of aspirin in each tablet.
STEP 3
Translate it into maths.
(450 × 3 × 4) ÷ 300 =
STEP 4
Estimate the answer:
• Simplify the numbers by rounding off to make the calculations easier.
• Use your common sense/clinical judgement to make a logical guess.
First simplify 450 multiplied by 4
400 × 4 = 1600
50 × 4 = 200
1600 + 200 = 1800
1800 × 3 = 5400
5400 divided by 300 can be simplified to 54 ÷ 3 = 18
STEP 5
Do the precise calculation using a calculator or pen and paper.
450 × 3 = 1350
1350 × 4 = 5400
5400 ÷ 300 = 18
STEP 6
Interpret the answer and put it into words.
You would need eighteen tablets for the whole course of treatment, which means one bottle is plenty.
STEP 7
Check the answer.
If you got it wrong, decide what skills/knowledge you need to revise.
• Choosing the right formula.
• Estimating in your head by simplifying calculations.
• Using a calculator or pen and paper to do divisions and multiplication.
• Using a calculator or pen and paper to do divisions and multiplications.

Question 3

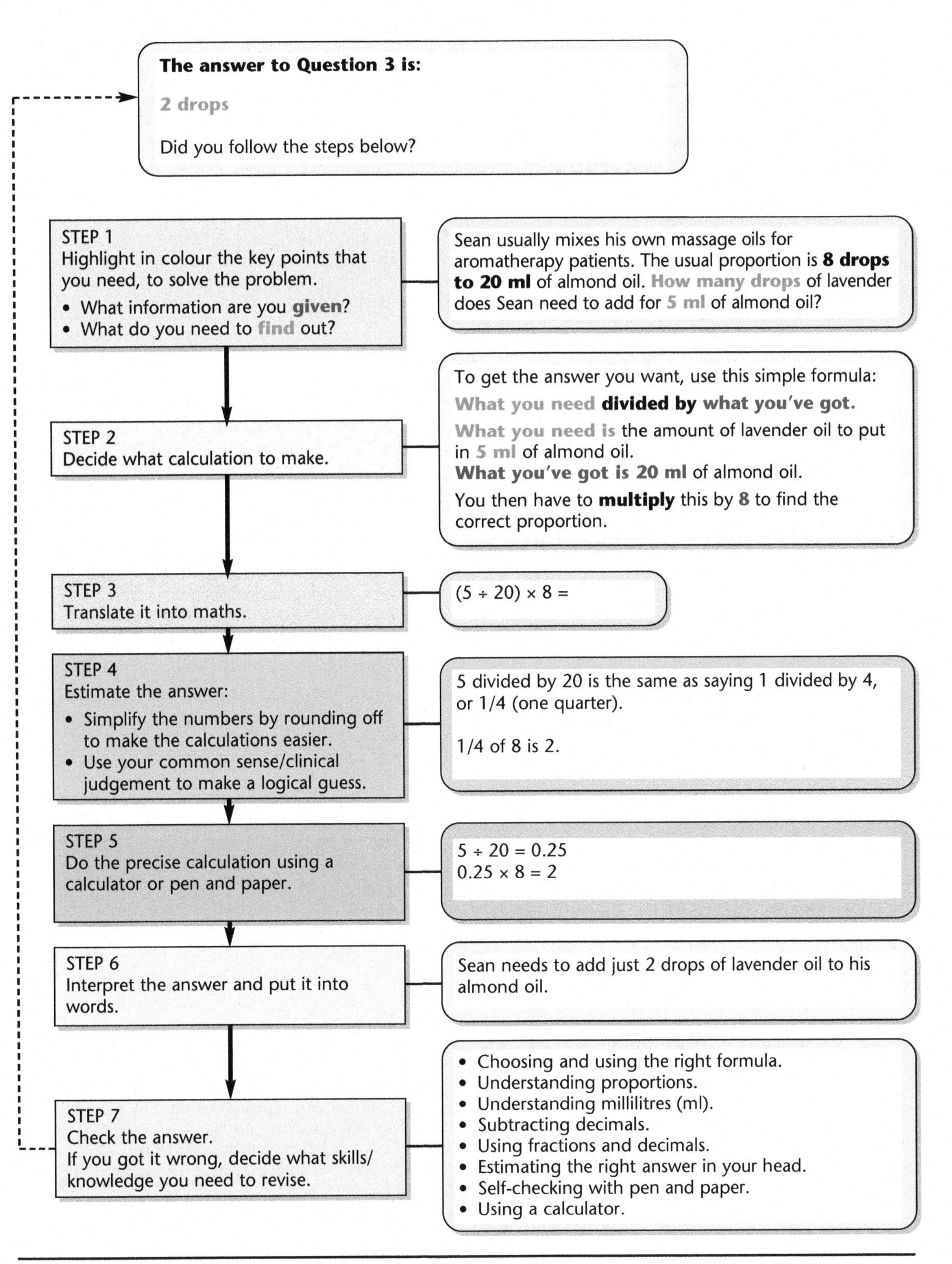

Diagram created in Inspiration® by Inspiration Software®, Inc.

Question 4

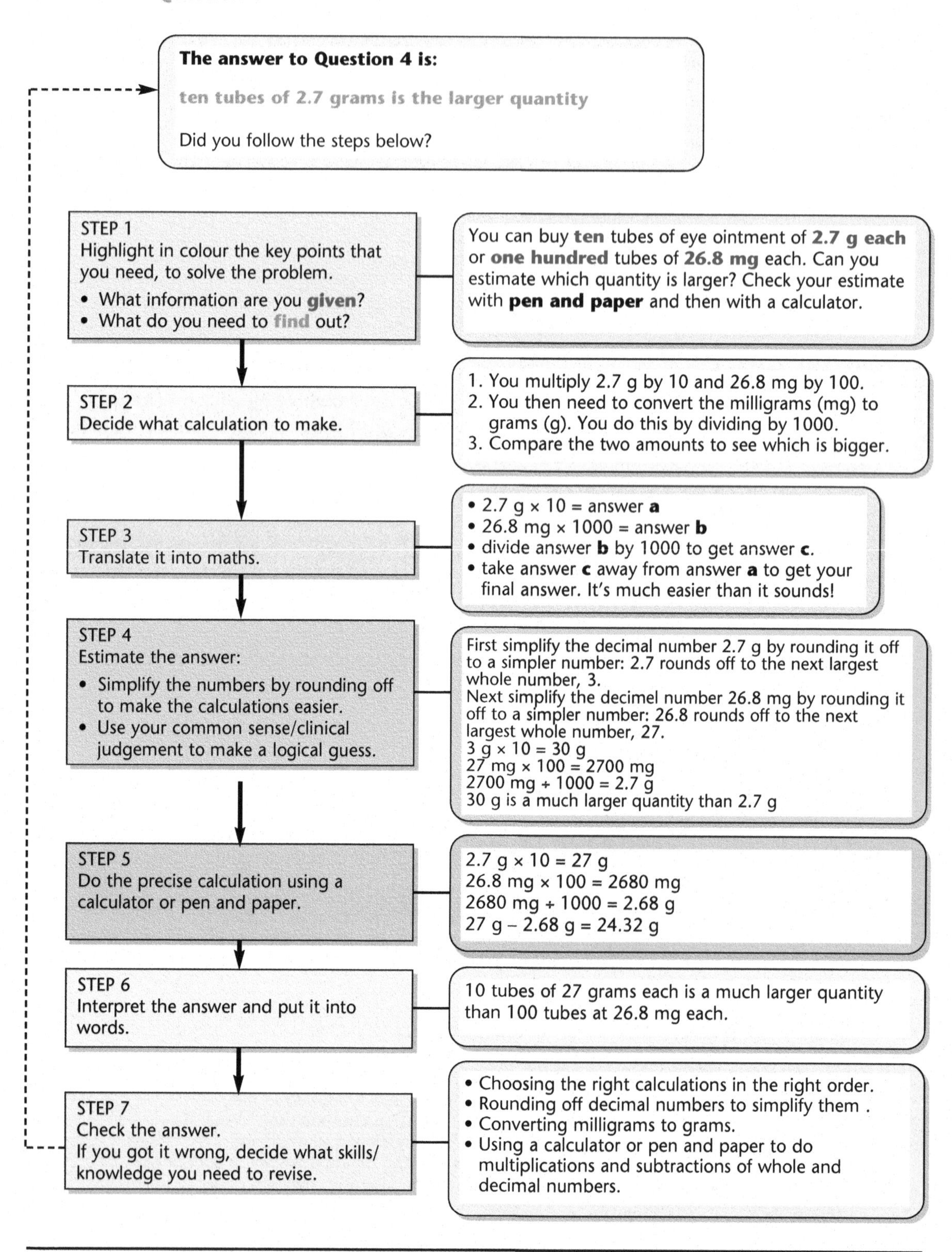

Diagram created in Inspiration® by Inspiration Software®, Inc.

Question 5

The answer to Question 5 is:

The answer to Question 5 is found in Table 2, below. You need to use your logic for this one. Adding minutes and hours can be confusing. It really helps to draw up a table so that you can see clearly what you are doing.

An osteopath's practice opens at 9 a.m. Anne has an appointment at **9 for 30 minutes**. John at **9.30 for 20** minutes. Moira at **9.50 for 5** minutes. **Ali at 9.55** for 10 minutes. **Moira and Ali** have to get to work **as early as possible**. How could the appointments be re-scheduled so that **waiting times are reduced**?

Table 1: THE ORIGINAL APPOINTMENTS

Client	Appointment time	Length of appointment
Anne	9	30 mins
John	9.30	20 mins
Moira	9.50	5 mins
Ali	9.55	10 mins

Table 2: THE NEW APPOINTMENTS

Client	Appointment time	Length of appointment
Moira	9	5 mins
Ali	9.05	10 mins
John	9.15	20 mins
Anne	9.35	30 mins

Question 6

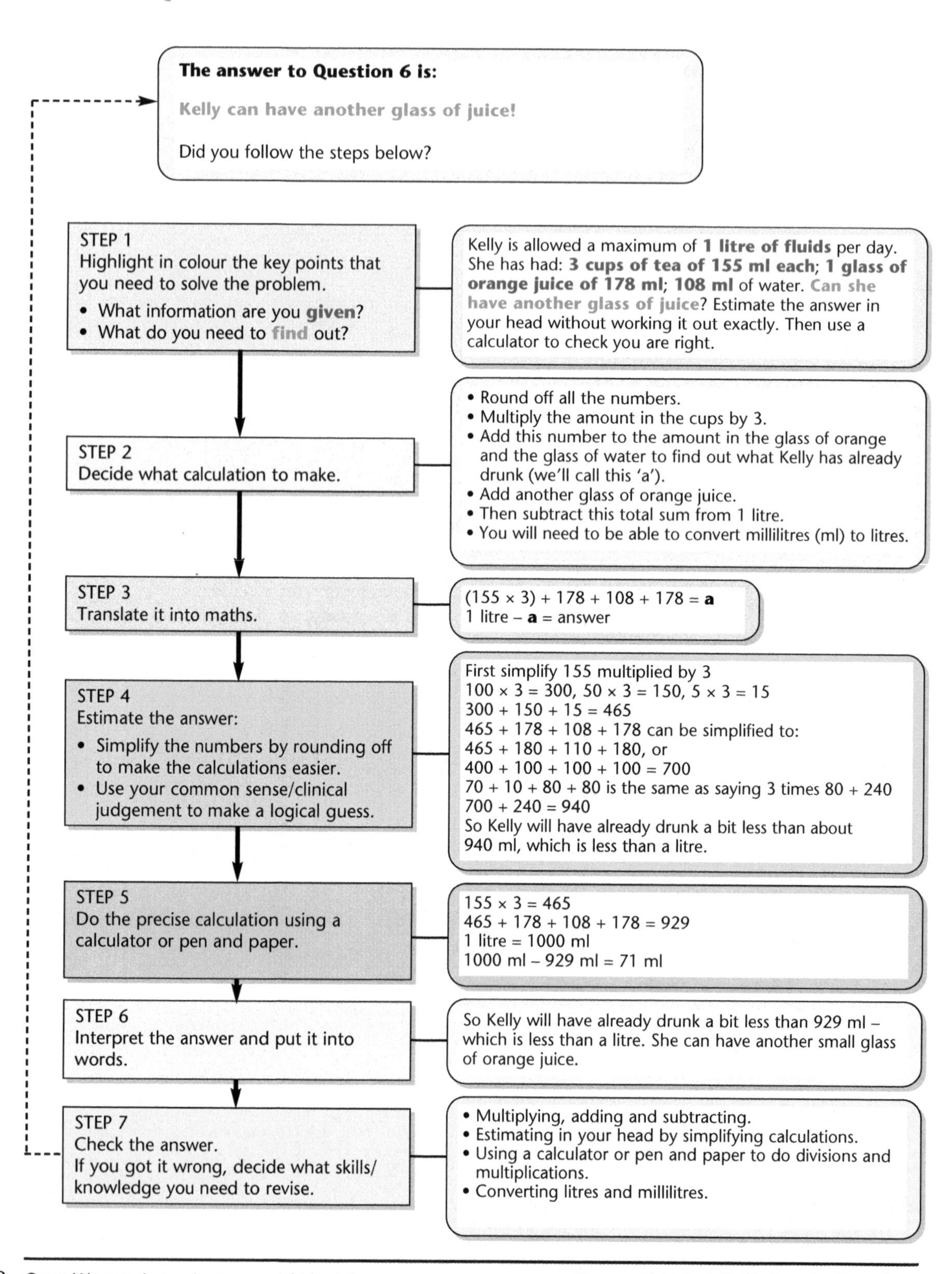

Diagram created in Inspiration® by Inspiration Software®, Inc.

Question 7

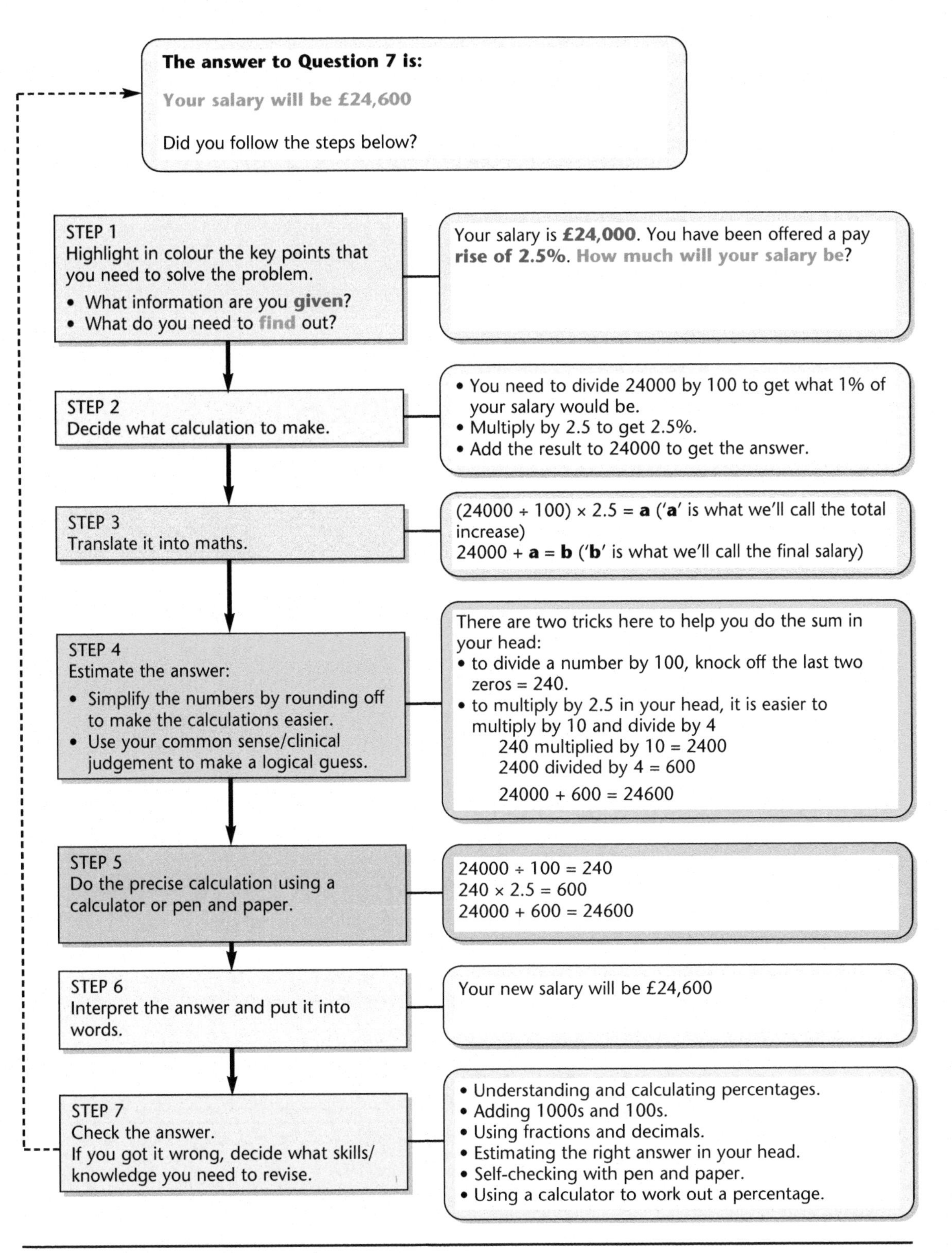

Diagram created in Inspiration® by Inspiration Software®, Inc.

Question 8

The answer to Question 8 is:

32.2 (with a remainder of 22).

To calculate Sally's BMI, you divide 77.5 by 24 (to find the square of her height, multiply 1.55 × 1.55 = 2.4).

There are no less than **10** steps you need to take if you use pen and paper to work out the LONG DIVISION needed to answer question 8!

The answer is 32.2.

Below you will find an explanation for each step.

No wonder human beings invented the pocket calculator!

```
   32.2
24 775
  -72
    55
   -48
    70
   -48
    22
```

1 First simplify the numbers you have to work with, by converting the decimal numbers to whole numbers: **77.5** divided by **2.4** is the same as saying **775** divided by **24**.
You now have much simpler numbers to work with!

2 Set out the calculation that you need to make, like this:

24 **775**

On the left is the number you divide by (this number is called the 'divider'). On the right is the number you are dividing up.

3 Ask yourself 'How many times does **24** go into **77**?
The answer is **3** times, so put the **3** above the **775**.

4 Now you need to find out if there is a 'remainder' or number left over when you divide **77** by **24**, so you take away **72** (**3** times **24**). The 'remainder' (what's left over) is **5**.

5 Then you drop down the next number from the original number you are dividing up (**775**) and now you have a new number (**55**) to divide by **24**.
And you start all over again

6 Ask yourself 'How many times does **24** go into **55**?'
The answer is **2** times, so put the **2** above the **775**.

7 Now you need to find out if there is a 'remainder' or number left over when you divide **55** by **24**, so you take away **48** (**2** times **24**).
The 'remainder' (what's left over) is **7**.

8 If you want to calculate your answer a bit more precisely, you carry on to one decimal point after the whole number. As there are no more numbers to drop down, we drop down **0** and now you have a new number (**70**) to divide by **24**).

9 Ask yourself 'How many times does **24** go into **70**?'
The answer is **2** times, so put the **2** above the **775**.

10 Now you need to find out if there is a 'remainder' or number left over when you divide **70** by 24, so you take away **48** (**2** times **24**) from **70**: this gives you **22**.

Your answer is **32.2** with a remainder of **22**.

Question 9

The answer to Question 9 is:

Sally's weight *is* a cause for concern.
See why below.

From Question 8, you will know that Sally's Body Mass Index (BMI) is 32.2. This means that her BMI is just over 32 and less than 33.

Look at the table relating to BMI below. If you look down the first column, you will find an entry that says 30–39.9 This refers to **the range of numbers** between 30 and 39.9 (just under 40). This means that, if a person's BMI falls within this range, she is unfortunately classified as 'obese':

Body Mass Index Table

Below 20	underweight
20–24.9	ideal
25–29.9	overweight
30–39.9 →	obese

It is so easy to miss out a line or copy out the wrong number when you are reading information laid out in a 'table' form – especially if you are stressed or in a hurry. It really helps to use a ruler or piece of card to locate and mark the information you want (see above).

When we calculate somebody's BMI, we are comparing their weight with their height. This calculation gives you the **ratio** between weight and height. If a woman weighs 63 kilos (10 stone), but her height is 1 metre 20 (4 ft), then her BMI will be **high**. If a woman weighs 63 kilos (10 stone) and her height is 1 metre 80 (6 ft) then her BMI is **very low**. So what is important is the ratio between weight and height, not just a person's weight.

Question 10

The answer to Question 10 is:

Emma's pulse rate returns to her normal resting rate (60 beats per minute) 8 minutes after she has stopped exercising.

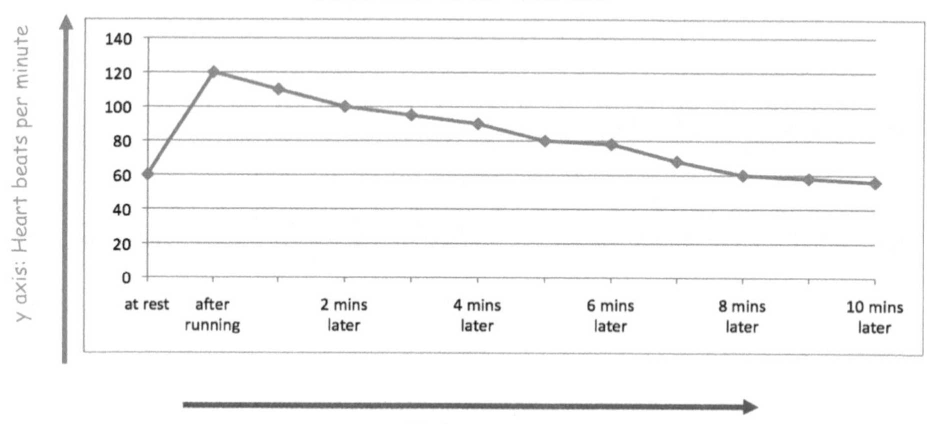

x axis: Elapsed time

The horizontal line ⟶ for the elapsed (passed) time is called the '***x* axis**'

The vertical line ↑ for the heart beats per minute (pulse rate) is called the '***y* axis**'

- Find 'at rest' on the ***x*** axis: now run your finger up the ***y*** axis until you meet the graph trace.
- Find the 'at rest' pulse rate from the ***y*** axis (60 heart beats per minute). Then run your finger parallel to the ***x*** axis until the graph trace crosses the 60 line again.
- Run your finger down the ***x*** axis and read the figure: this gives you your answer.

Accurately reading and entering figures on graphs can be difficult if you are hurried, stressed or working in a distracting environment. It is quite easy to make serious errors.

It helps to use a highlighter pen to mark lines across and up/down the list to make it easier to follow along a column of figures or read from a graph.

If you can't use a highlighter on a chart, use a set square:

This will help you to find your place across a sheet, both horizontally and vertically.

Handy tips for quick painless calculations

Multiplying and dividing decimal numbers by 10, 100 or 1000

1. Decimals numbers

The number 2.25 has 3 digits or figures.

For example:
2.25 pints of beer means two and a quarter pints:
a bit more than 2 whole pints, but much less than 3 pints.

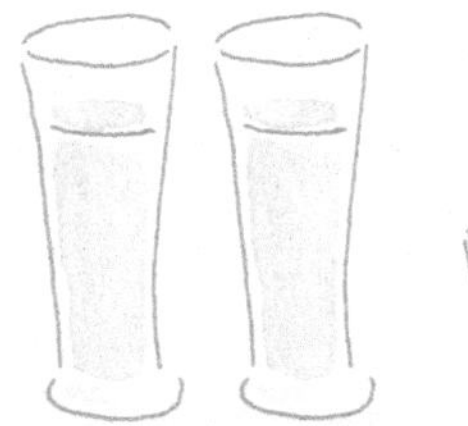

2. Is there an easy way to multiply a decimal number by 10, 100 and 1000?

You just move the decimal point to the right. See below:

If you multiply a decimal number by:	Number of places to move the decimal point to the right	Example
10 100 1000	1 2 3 (or add a zero, if there are no figures left to move along!)	2.25 × 10 = 22.5 2.25 × 100 = 225 2.25 × 1000 = 2250

3. Is there an easy way to divide a decimal number by 10, 100 and 1000?

If you divide a decimal number by:	Number of places to move the decimal point to the left	Example
10 100 1000	1 2 3 (or add a zero, if there are no figures left to move along!)	225 ÷ 10 = 22.5 225 ÷ 100 = 2.25 225 ÷ 1000 = 0.225

4. Why is it important to be able to multiply or divide decimals quickly by 10, 100 and 1000?

Being able to multiply or divide decimals quickly by 10, 100 and 1000 allows you to convert between units of money, weight, size or liquid.

For example:
Suppose you need to work out a drug dose calculation and you have the formula

$$\frac{800\ \mu g}{1.5\ mg} \times 10$$

Before you start to make your division, you need to make sure that the units are the same. So you need to convert 1.5 mg to μg by multiplying by 1000. If you make an error in placing the decimal point, this could have very serious consequences for a patient.

You will find some top tips and examples for conversions over the next few pages.

Some examples of easy conversions

Money

To convert pounds into pence, multiply by 100

Pounds	Pence	How we get the answer
£1	= 100 p	Multiple 1 by 100
£1.55	= 155 p	Multiply 1.55 by 100

To convert pence into pounds, divide by 100

Pence	Pounds	How we get the answer
125p	= £1.25	Divide 125 by 100
1550p	= £15.50	Divide 1550 by 100

Size

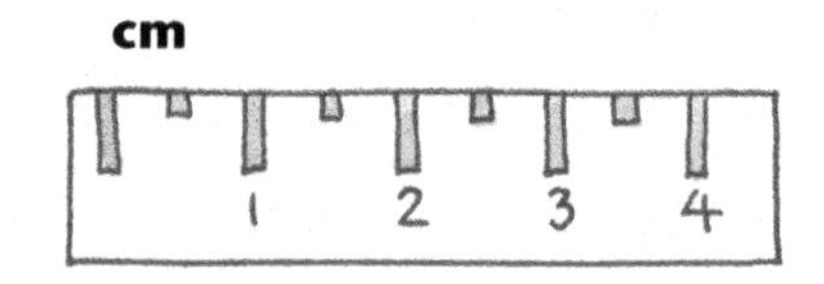

Metre is abbreviated to: m
Centimetres is abbreviated to: cm
Millimetre is abbreviated to: mm

Remember: 1 centimetre (cm) = 10 millimetres (mm)

To change centimetres into millimetres, multiply by 10

Centimetres	Millimetres	How we get the answer
1 cm	= 10 mm	Multiply 1 by 10
1.25 cm	= 12.5 mm	Multiply 1.25 by 10

To change millimetres into centimetres divide by 10

Millimetres	Centimetres	How we get the answer
20 mm	= 2 cm	Divide 20 by 10
25 mm	= 2.5 cm	Divide 25 by 10

***Remember*:** 1 metre (**m**) = 100 centimetres (**cm**)

To change metres into centimetres, multiply by 100

Metres	Centimetres	How we get the answer
1 m	= 100 cm	Multiply 1 by 100
1.25 m	= 125 cm	Multiply 1.25 by 100

To change centimetres into metres divide by 100

Centimetres	Metres	How we get the answer
5300 cm	= 53 m	Divide 5300 by 100
2000 cm	= 20 m	Divide 2000 by 100

Weight

Kilogram is abbreviated to: **kilo** or **kg**
Gram is abbreviated to: **g**

To change kilograms into grams, multiply by 1000

Kilos	Grams	How we get the answer
1 kg	= 1000 g	Multiply 1 by 1000
1.25 kg	= 1250 g	Multiply 1.25 by 1000

To change grams into kilograms, divide by 1000

Grams	Kilos	How we get the answer
5300 g	= 5.3 kg	Divide 5300 by 1000
2000 g	= 2 kg	Divide 2000 by 1000

Gram is abbreviated to: g
Milligram is abbreviated to: mg

To change grams into milligrams, multiply by 1000

Grams	Milligrams	How we get the answer
1 g	= 1000 mg	Multiply 1 by 1000
1.25 g	= 1250 mg	Multiply 1.25 by 1000

To change milligrams into grams, divide by 1000

Milligrams	Grams	How we get the answer
5300 mg	= 5.3 g	Divide 5300 by 1000
2000 mg	= 2 g	Divide 2000 by 1000

Milligram is abbreviated to: mg
Microgram is abbreviated to: mcg or µg

To change milligrams into micrograms, multiply by 1000

Milligrams	Micrograms	How we get the answer
1 mg	= 1000 µg	Multiply 1 by 1000
1.25 mg	= 1250 µg	Multiply 1.25 by 1000

To change micrograms into milligrams, divide by 1000

Micrograms	Milligrams	How we get the answer
5300 µg	= 5.3 mg	Divide 5300 by 1000
2000 µg	= 2 mg	Divide 2000 by 1000

Liquid

Litre is abbreviated to: **l**
Millilitre is abbreviated to: **ml**

To change litres into millilitres, multiply by 1000

Litres	Millilitres	How we get the answer
1 l	= 1000 ml	Multiply 1 by 1000
1.25 l	= 1250 ml	Multiply 1.25 by 1000

To change millilitres into litres divide by 1000

Millilitres	Litres	How we get the answer
5300 ml	= 5.3 l	Divide 5300 by 1000
2000 ml	= 2 l	Divide 2000 by 1000

Maths Emergency (ME) notebook

Many of the calculations that you will need will crop up again and again. Many people have found that keeping a small notebook in their pocket for maths emergencies is a life-saver – literally in some cases, as even small errors in calculations could be a serious health hazard.

Your notebook will be personal to you only, containing the tips, examples, short cuts and formulae you may need for reference. If you have to make drip calculations, you may also find it very useful to carry a calculator with common drip rates on a ready reckoner taped to it.

Over the next few pages, you will find some ideas of useful things to put in your ME notebook.

Multiplying and dividing decimals quick calculator

When you multiply a decimal number . . .	move the decimal point . . .
by 10 by 100 by 1000	1 space to the right 2 spaces to the right 3 spaces to the right
When you divide a decimal number . . .	**move the decimal point . . .**
by 10 by 100 by 1000	1 space to the left 2 spaces to the left 3 spaces to the left

Quick calculator for conversions

LIQUID:	**To change litres into millilitres** **To change millilitres into litres**	**multiply** **divide**	**by 1000** **by 1000**
SIZE:	To change metres into centimetres To change centimetres into metres	multiply divide	by 100 by 100
	To change centimetres into millimetres To change millimetres into centimetres	multiply divide	by 10 by 10
WEIGHT:	To change kilograms into grams To change grams into kilograms	multiply divide	by 1000 by 1000
	To change grams into milligrams To change milligrams into grams	multiply divide	by 1000 by 1000
	To change milligrams into micrograms To change micrograms into milligrams	multiply divide	by 1000 by 1000

Basic formula used to calculate drug dosages

What to GIVE =	
• **What you NEED** • what you HAVE • what FORM it's in	divided by multiplied by (tablets, liquid, amount etc.)

Useful resources

In this section you will find a short list of resources that other students have found very useful. The addresses for websites may change, but details were correct when this book was published.

General study skills

The Study Skills Handbook by Stella Cottrell (Palgrave Macmillan, 2013)

Calculations

Calculations for Nursing and Healthcare by Diana Coben and Elizabeth Atere-Roberts (Palgrave Macmillan, 2005)

Dosagehelp www.dosagehelp.com
Tutorials examples and practice tasks for calculating dosages

Skillswise Maths www.bbc.co.uk/skillswise/maths
Practical common sense maths for adults

3D and audio-visual resources

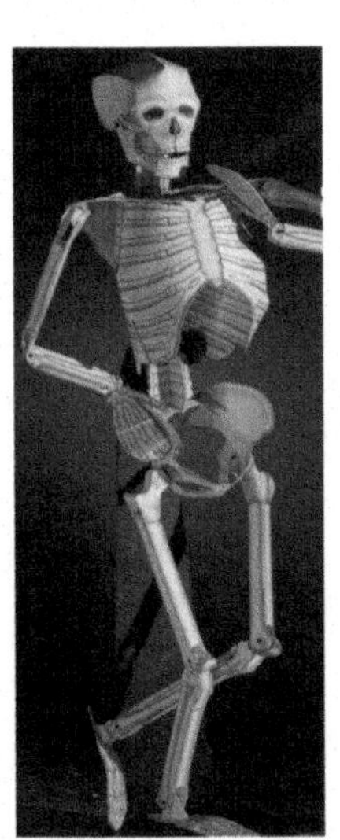

Cut and Make a Human Skeleton by A.G. Smith (Dover Publications, 2013) www.doverpublications.com

You have to be quite rich to afford a real skeleton! An excellent alternative is to make a cardboard version. If you are a practical hands-on kind of person, you will probably find this far more useful than a diagram of a flat skeleton.

Human Anatomy Atlas by Visible Body

A body visualization app with over 3000 structures from all systems in the human body.

Mental Wellbeing

Get Some Headspace: 10 Minutes Can Make All the Difference by Andy Puddicombe (Hodder Paperbacks, 2012)

http://www.getsomeheadspace.com free online meditation with a contemporary feel

Mindfulness: A practical guide to finding peace in a frantic world by Prof Mark Williams and Dr Danny Penman (Piatkus, 2011)

Mind http://www.mind.org.uk
Helpline, information and advice covering a wide range of mental health issues.

Technology

Medical spellcheckers, dictionaries and word lists

Handheld spellcheckers and spellchecking software
Spellex PocketMed Spellex www.spellex.com
Hand Held Medical Spell Checker

Medilexicon www.medilexicon.com
Free medical dictionaries with abbreviations and definitions.

Word lists
Wordbar www.cricksoft.com
A talking word bank tool giving you rapid access to personal lists of useful words that you have stored. You also get access to hundreds of ready-made Wordbars available from free online resources.

Visual Thesaurus
http://www.visualthesaurus.com
Creates an animated webbed diagram with your word in the centre of the display, connected to related words and meanings. It includes a spell checker, word suggestion, and text-to-speech.

Aids to reading

Scanning Pens with medical dictionary included
Quicktionary 2 Premium Professional

Turning text into speech
Read & Write Gold: from Texthelp www.texthelp.com/uk

Aids to note-taking

Audio Notetaker from Sonocent www.sonocent.com/en/audio_notetaker
Software to help you review and organize key sections of your recordings.

Aids to writing

Speech Recognition – turns speech into written text
Dragon Dictate Medical: from Nuance http://www.nuance.com/index.htm

Mind Mapping software

Inspiration: www.inspiration.com

MindView: from MatchWare www.matchware.com/en/products/mindview

MindManager: from Mindjet www.mindjet.com/uk

For brainstorming, planning, noting, organizing, reviewing, memorizing and presenting information in map or chart format.

Memory tools

Quizlet www.quizlet.com
Free software to help you learn any subject, including a vast database of flashcards. Includes facilities for tracking your progress and testing yourself, as well as game modes to help you enjoy memorizing your subject.

Eclipse Crossword www.eclipsecrossword.com
A fast, easy, free way to create crossword puzzles for revision purposes.

Answers

Answers to Questionnaire: Exploring myths about exams

Score one point if you have ticked 1b, 2b, 3a, 4a, 5a, 6a, 7b.

If you have 6 or 7 points: your approach to exams seems positive and realistic.

If you have 4 points or 5 points: your expectations of yourself are fairly positive, but you may still be causing yourself some unnecessary stress.

If you have less than 4 points: your expectations of yourself seem unrealistic – this could make you really stressed and interfere with performance. Changing some of these thought and behaviour patterns may help you get rid of negative feelings and stress. You may also like to try out some of the other stress-busting strategies presented in this chapter.

Answers to Activity: Linking words

Hydrophilic: water loving
Haemophiliac: blood disorder due to deficiency of coagulation factor
Bicuspid point: double pointed
Erythrocyte: red blood cell
Cytokinesis: cell movement
Cardiovascular: involving the heart and the blood vessels
Endoscopy: internal examination
Telophase: last phase of cell division
Prophase: first stage of cell division

Solutions to the crossword puzzles

The circulatory system – answers

Clues Across

2. **MYOCARDIUM** – This is the middle layer of the muscular wall of the heart and is made of muscular tissue.

5. **GRANULOCYTES** – These make up 75% of the leucocytes and defend the body against viruses and bacteria.
6. **VENTRICLES** – These are the bottom left and right chambers of the heart.
9. **LYMPHOCYTES** – These are formed in lymphatic tissue and produce antibodies.
10. **PERICARDIUM** – This covers the outside of the heart.
12. **PHAGOCYTOSIS** – The process by which bacteria are devoured.
13. **ATRIA** – These are the top right and left chambers of the heart.
14. **SYSTEMIC** – This is the name of the circulation from the heart to the body.
15. **PLASMA** – This makes up 55% of the blood and helps transport essential substances round the body.

Clues Down

1. **BICUSPID** – Blood is pushed by contraction of the left atrium into the left ventricle through this valve.
3. **THROMBOCYTES** – These are a type of blood cell that help with blood clotting.
4. **LEUCOCYTES** – These protect the body from infection and increase by mitosis when the body is attacked.
7. **ENDOCARDIUM** – This is the inner layer of the heart's muscular wall.
8. **ERYTHROCYTES** – These are also known as red blood cells.
11. **PULSE** – The rate at which your heart pumps blood through the circulatory system.

The digestive system – answers

Clues Across

1. **EPIGLOTTIS** – This prevents choking.
4. **FAECES** – The unpleasant unwanted products of food.
9. **TONGUE** – This contains taste buds.
10. **SALIVA** – This is secreted by glands and helps start digestion.
11. **PANCREATIC** – These juices help digest food in the small intestine.
12. **PEPSIN** – Enzyme that turns proteins into amino acids.

Clues Down

2. **LIVER** – Sugars and amino acids are passed to this organ.
3. **ABSORPTION** – This is when digested food is absorbed through the walls of the villi.
5. **ENZYMES** – These make food more digestible.
6. **PERISTALTIC** – This type of movement pushes food against the villi.
7. **DIGESTION** – Breakdown and transformation of food into substances transported by the blood.
8. **STOMACH** – Place where protein is digested.

The endocrine system – answers

Clues Across

4. **HYPERSECRETION** – This means that too much of a hormone is produced.
5. **PANCREAS** – This organ secretes insulin in the islets of Langerhans.

6. **THYMUS** – This gland is part of the immune system.
10. **THYROID** – Hormones secreted by these glands stimulate growth.
11. **HORMONES** – These are chemical messengers produced by glands.
12. **OESTROGEN** – The ovaries secrete this as well as progesterone.
14. **PARATHYROID** – This kind of gland is situated on both sides of the thyroid gland.
15. **ADRENALINE** – This hormone is released to help us fight a threat or run away from it.

Clues Down

1. **MELATONIN** – The pineal gland secretes this.
2. **MEDULLA** – The inner part of the adrenal gland.
3. **TESTOSTERONE** – The testes secrete this.
5. **PITUITARY** – This kind of gland is situated at the base of the brain.
7. **HYPOGLYCEMIA** – This occurs when there is a lower than normal level of blood glucose.
8. **CIRCADIAN** – This kind of rhythm means that regular fluctuations of hormone levels occur every 24 hours.
9. **HOMEOSTASIS** – This means that the internal environment of the body is kept in balance.
13. **DUCTLESS** – A kind of gland which has no separate tube or canal to carry hormones into the bloodstream.

The lymphatic system – answers

Clues Across

1. **LYMPHOCYTES** – Lymph contains waste materials such as these.
4. **INGUINAL** – These nodes are situated near the genital area.
6. **LEUCOCYTES** – Lymph contains waste materials such as these.
9. **LYMPH** – This fluid is similar to blood plasma and is filtered by the lymphatic nodes.
11. **PHAGOCYTES** – Lymphatic tissue contains cells such as these.
12. **TONSILS** – These contain lymphatic tissue and are situated in the throat.
13. **AXILLARY** – These nodes are situated near the armpits.

Clues Down

2. **THYMUS** – This gland contains lymphatic tissue and is situated behind the sternum.
3. **SPLEEN** – This is a non-essential organ which produces and destroys cells.
5. **NODES** – The lymph vessels open into these structures, which can be found throughout the body.
7. **OCCIPITAL** – These nodes are situated at the back of the skull.
8. **APPENDIX** – This non-essential organ contains lymphatic tissue and is part of the digestive system.
10. **OEDEMA** – This is the swelling of tissues, caused by an obstruction to the lymphatic flow.

The nervous system – answers

Clues Across

5. **EPINEPHRINE** – Stress hormone
6. **SYMPATHETIC** – The division of the Autonomic Nervous System that prepares the body for fight or flight.
8. **MENINGES** – Membranes which protect the Central Nervous System.
10. **ADRENAL** – Glands where stress hormones are produced.
12. **LIMBIC** – System in charge of emotions.
15. **PARASYMPATHETIC** – This system restores the body to a resting state.
16. **CORTEX** – Outer layer of the brain.

Clues Down

1. **HYPOTHALAMUS** – This is the size of a pea and controls body temperature.
2. **PONS** – Joins the hemispheres of the cerebellum and connects the cerebrum to the cerebellum.
3. **CEREBELLUM** – Regulates balance, movement and muscle coordination.
4. **NEURONE** – Another word for nerve cell.
7. **MYELIN** – Sheath covering the axon.
9. **DENDRITES** – Nerve fibres which transmit impulses to the cell body.
11. **PITUITARY** – Gland at base of brain which secretes hormones.
13. **SYNAPSE** – Where one neurone meets another.
14. **NUCLEUS** – Centre of a nerve cell.

The reproductive system – answers

Clues Across

1. **PLACENTA** – The unborn baby's support system.
4. **OVA** – The eggs that the female reproductive system produces.
7. **VAGINA** – This muscular passage connects the cervix to the vulva.
8. **POLYCYSTIC** – A type of ovarian syndrome.
10. **UTERUS** – The fertilized ovum grows into a baby here.
11. **ECTOPIC** – A pregnancy that occurs outside the uterus.
13. **SPERMATOZOA** – These are produced and stored in the testes.
15. **SCROTUM** – The sac of skin and muscle where the testes are contained.
17. **SEMEN** – The fluid ejaculated during intercourse.
18. **OVARIES** – Glands that secrete oestrogen and progesterone.
19. **EPIDIDYMIS** – A tightly coiled tube which transports sperm.

Clues Down

1. **PROSTATE** – A small gland between the bladder and rectum in men.
2. **AMNIOTIC** – This fluid protects the unborn baby from shocks.
3. **ZYGOTE** – The first cell of the baby.
5. **FALLOPIAN** – The uterus connects to these tubes.
6. **DYSMENORRHOEA** – Very painful menstruation.
9. **CHROMOSOMES** – The nucleus in the head of a sperm contains 23 of these.
12. **TESTES** – Glands contained within the scrotum.

14. **FORESKIN** – This piece of skin is sometimes removed from the penis for religious or hygienic reasons.
16. **CERVIX** – The neck of the womb.

The skeletal system – answers

Clues Across

7. **PERIOSTEUM** – White skin-like covering of most bones.
8. **LONG** – Type of bone that allows the movement of the limbs.
9. **FLAT** – Bones with broad surfaces for muscle attachment, such as the shoulder blades.
12. **MARROW** – This is where fats are stored in the bone.
13. **CANCELLOUS** – This type of tissue looks like a sponge.

Clues Down

1. **SESAMOID** – The type of bone that makes up the knee-cap.
2. **IRREGULAR** – All the bones of the face are like this.
3. **HAVERSIAN** – These canals run through the bone tissue.
4. **AXIAL** – This skeleton supports the head, neck and torso.
5. **APPENDICULAR** – This type of skeleton supports the limbs.
6. **OSTEOBLASTS** – Bones are made of these cells.
10. **SHORT** – Strong bones but where little movement is required.
11. **BONES** – 206 of these make up the skeleton.

The skin – answers

Clues Across

4. **APOCRINE** – These sweat glands produce a fluid which creates body odour when mixed with bacteria on the skin surface.
5. **CORNEUM** – The top surface of the epidermis.
7. **DERMIS** – Lies beneath epidermis and contains sweat glands.
10. **MAST** – These cells produce histamine in response to an allergic reaction.
12. **PAPILLA** – These conical projections contain blood vessels and nerves which supply the hair with nutrients.
13. **SEBACEOUS** – These glands produce sebum.
14. **GERMINATIVUM** – Bottom layer of epidermis, containing melanin pigment of the skin.

Clues Down

1. **ECCRINE** – These sweat glands help control body temperature.
2. **LEUCOCYTES** – These are white blood cells that fight disease and infection.
3. **LUCIDUM** – The second clear layer of the epidermis.
6. **FIBROBLASTS** – These are cells responsible for the production of collagen, elastin and areolar tissue.
8. **GRANULOSUM** – The third, granular layer of the epidermis.
9. **FOLLICLES** – These contain hair.

11. **ELASTIN** – The dermis contains collagen and this kind of tissue which keeps the skin supple and elastic.

The urinary system – answers

Clues Across

3. **FILTRATION** – This takes place in the Bowman's capsule.
4. **MEDULLA** – This is the inside part of the kidney.
5. **CORTEX** – This is the outside of the kidney.
7. **UREA** – Urine contains about 2% of this substance.
8. **BLADDER** – This holds urine.
9. **SPHINCTER** – This relaxes when the bladder wall contracts, emptying the urine into the urethra.
10. **KIDNEYS** – Bean-shaped organs that filter toxic substances from blood.

Clues Down

1. **GLOMERULUS** – Capillaries surrounded by the Bowman's capsule.
2. **URETHRA** – Tube from bladder to outside of body. It is the passage for semen in men.
6. **URETERS** – Tubes connecting kidneys to bladder.

Index

·y Printforce, the Netherlands